Alfred Binet

The Psychic Life of Micro-Organisms

a study in experimental psychology

Alfred Binet

The Psychic Life of Micro-Organisms
a study in experimental psychology

ISBN/EAN: 9783337332549

Printed in Europe, USA, Canada, Australia, Japan

Cover: Foto ©Thomas Meinert / pixelio.de

More available books at **www.hansebooks.com**

THE PSYCHIC LIFE

OF

MICRO-ORGANISMS

A STUDY IN EXPERIMENTAL PSYCHOLOGY

BY

ALFRED BINET

AUTHORISED TRANSLATION

CHICAGO
THE OPEN COURT PUBLISHING COMPANY
1894.

PREFACE TO THE AMERICAN EDITION.

I HAVE endeavored, in the following essay upon Micro-organisms, to show that psychological phenomena begin among the very lowest classes of beings; they are met with in every form of life from the simplest cellule to the most complicated organism. It is they that are the essential phenomena of life, inherent in all protoplasm.

We admit, accordingly, the existence of a vitalism, that is to say, of an aggregate of properties which properly pertain to living matter and which are never found in inanimate substances. Among these properties of life we classify psychological phenomena.

Vitalism, it is unnecessary to say, has nothing in common with the doctrine upheld by the School of Montpellier. The principle here involved has nothing to do with properties and forces that are superadded to living matter; it concerns the properties that are inherent in it—the properties that characterize life.

The modern opponents of vitalism seek to confute the theory by attempting to explain all phenomena of life from physico-chemical forces. They maintain that according as physiology advances the tendency is to relegate all phenomena nominally physiological into the domain of physics and chemistry; and that it would be only a question of time, if as yet they had not succeeded in demonstrating that every vital process is founded upon mechanical phenomena.

In a recent treatise upon "Vitalism and Mechanism,"* M. Bunge, professor of physiology at Basel, has shown that the history of physiology disproves these hypotheses. The more closely

---

* G. Bunge, *Vitalismus und Mechanismus*, Ein Vortrag, 1886.

the phenomena of life are scrutinized, the more carefully they are studied in their various aspects, the more certain does the conclusion become that the processes attributed to physico-chemical forces in reality obey much more complicated laws. To illustrate, it was at one time conceded that the phenomena of resorption and nutrition were explainable by diffusion and endosmosis; Dutrochet, upon his discovery of endosmosis, imagined even that he had discovered the principle of life. At the present time we know that the walls of the intestines do not in any wise act like the inanimate membrane used in experiments in endosmosis. They are covered with epithelial cells, each of which is an organism endowed with a complex of properties. The protoplasm of these cells lays hold of food by an act of prehension, exactly as the ciliate Infusoria and other unicellular organisms do, that lead an independent life. In the intestines of cold-blooded animals the cells emit prolongations wnich seize the minute drops of fatty matter and, carrying them into the protoplasm of the cell, convey them thence into the chylifactive ducts. There is still another mode of absorption of fatty matters, met with among cold-blooded as well as warm-blooded animals: the lymphatic cells pass out from the adenoid tissue which contains them, so that upon arriving at the surface of the intestines they seize the particles of fatty matter there present and, laden with their prey, make their way back to the lymphatics.

Accordingly, the faculty of seizing food and of exercising a choice among foods of different kinds—a property essentially psychological—appertains to the anatomical elements of the tissues, just as it does to all unicellular beings, in the manner shown in our treatise. It is plainly impossible to explain these facts by the introduction of physico-chemical forces. They are the essential phenomena of life and are the exclusive appurtenance of living protoplasm.

If the existence of psychological phenomena in lower organisms is denied, it will be necessary to assume that these phenomena can be superadded in the course of evolution, in proportion as an organism grows more perfect and complex. Nothing could be more inconsistent with the teachings of general physiology, which shows us that all vital phenomena are previously present in non-differentiated cells.

Furthermore, it is interesting to note to what conclusion the admission would lead—as Romanes apparently does admit—that psychological properties are wanting in lower-class beings and that they enter at different stages of zoölogical development. Romanes has minutely particularized on a large chart the development of the intellectual powers, in quite an arbitrary manner. According to his scheme, only protoplasmic movements, and the property of excitability are present in lower-class organisms. Memory begins first with the echinoderms; the primary instincts with the larvæ of insects and the Annelids; the secondary instincts, with insects and spiders; reason, finally, commences with the higher Crustaceans.

I do not hesitate to say that all this laborious classification is artificial in the extreme, and perfectly anomalous.

All writers that have devoted themselves, with any pretension to special investigation, to the study of unicellular organisms, have attributed to these beings most of the psychological properties which M. Romanes reserves for this or that higher-class animal. This is the opinion of Gruber, of Verworn, of Mœbius, of Balbiani, and of many other naturalists. Mœbius recognizes that psychological life begins with living protoplasm, and he considers it to be the highest aim of zoölogy to demonstrate the psychical unity of all animals.

We could, if it were necessary, take every single one of the psychical faculties which M. Romanes reserves for animals more or less advanced on the zoölogical scale, and show that the *greater part* of these faculties belonged equally to Micro-organisms. But we must not unnecessarily extend the discussions of this introduction. We shall accordingly limit ourselves to few illustrations.

M. Romanes, in his zoölogical scale, assigns the first manifestations of surprise and fear to the larvæ of insects and to the Annelids. We may reply upon this point, that there is not a single ciliate Infusory that cannot be frightened, and that does not manifest its fear by a rapid flight through the liquid of the preparation.

If a drop of acetic acid be introduced beneath the glass-slide, in a preparation containing quantities of Infusoria, the latter will at once be seen to flee from all directions like a flock of frightened sheep.

Memory, according to M. Romanes, first begins with the Echinoderms. Now, Mœbius, upon the occasion of a treatise upon the *Folliculina ampulla*,* a ciliated Infusory presenting complicated and interesting movements, properly remarks that every time an animal repeats the same action under influence of the same excitations, that fact proves that the animal is possessed of memory. In fact, memory is one of the most elementary of psychological facts.

Lastly, the primary instincts, according to M. Romanes, begin first with the larvæ of insects and with Annelids. We give, in contradiction of this statement, the recent observations of Verworn,† which reveal the existence of curious instincts among the Rhizopods. The *Difflugia urceolata*, which inhabits a shell formed of particles of sand, emits long pseudopodia which search at the bottom of the water for the materials necessary to construct a new case for the filial organism to which it gives birth by division. The pseudopod, after having touched a particle of sand, contracts, and the grain of sand, adhering to the pseudopod, is seen to pass into the body of the animal.. Verworn, instead of grains of sand, placed small fragments of colored glass about the animal; some time afterwards, he noticed a heap of these fragments on the bottom of the shell. He then saw a bunch of protoplasm issue from the shell, representing the new *Difflugia* produced by division. Thereupon, the materials collected by the mother-organism—the fragments of colored glass—came forth from the shell and enveloped the body of the new individual in a sheath similar to that en-

---

* Mœbius, *Das Flaschenthierchen, Folliculina ampulla*, 1887.

† Verworn, *Zeitschrift für Wissenschaftliche Zoologie*, Bd. 46. H. 4. 1888.

casing the mother. These fragments of glass, loosely interjoined at first, were now cemented together by a substance secreted by the body of the animal.

Two facts are to be remarked in this observation: first, the act whereby the *Difflugia* collects the materials for providing the young individual with a case, is an act of preadaptation to an end not present, but remote; this act, therefore, has all the marks of an instinct. Further, the instinct of the *Difflugia* exhibits great precision; for the *Difflugia* not only knows how to distinguish, at the bottom of the water, the materials available for its purpose, but it takes only the quantity of material necessary to enable the young individual to acquire a well-built case; there is never an excess.

It is interesting to note that the *Difflugia* does not act differently from animals possessing more highly complicated organizations and endowed with differentiated nervous systems, as for instance, the larvæ of Phryganids which form their sheaths from shells, grains of sand, or minute slivers.

We shall not regard it as strange, perhaps, to find so complete a psychology in the history of lower organisms, when we call to mind that, agreeably to the ideas of evolution now accepted, a higher animal is nothing more than a colony of protozoans. Every one of the cells composing such an animal, has retained its primitive properties, giving them a higher degree of perfection by division of labor and by selection. The epithelial cells that secrete the nails and the hair are organisms perfected with reference to the secretion of protective parts. Similarly, the cells of the brain are organisms that have been perfected with reference to psychical attributes.

Paris, November 20, 1888.

        ‑   Alfred Binet.

TABLE OF CONTENTS.

VIII.

THE PHYSIOLOGICAL FUNCTION OF THE NUCLEUS.

IX.

CONCLUSION.

THE PSYCHIC LIFE OF MICRO-ORGANISMS.

THE study of microscopic organisms has hitherto been somewhat neglected by students of comparative psychology. Naturalists · who have devoted their attention to the study of these beings, have collected a great number of interesting facts concerning their psychic life; but these facts have not yet been critically examined and collated; they are scattered in reports and publications of all kinds, where the psychologist never dreams of looking for them. We shall endeavor to make him acquainted with a part of this wealth.

Under the name Micro-organism are included all those beings which by reason of their extreme smallness and simplicity of structure represent the lowest stages of animal or vegetable life; they constitute the very simplest forms of living matter, and consist of a single cell.

Some inhabit fresh and salt waters, serving as food for a great many other organisms, or contributing by means of their calcareous or silicious skeletons to the formation of continents. Others live as parasites in the organs of animals and plants, and induce more or less serious disorders in the constitutions of the organisms they have penetrated. Others, again, acting like ferments produce important chemical modifications in organic matter in the course of decomposition.

A great number of classifications for the methodical distribution of these beings has been proposed; but not one of them is altogether satisfactory; and that stands to reason. If a natural classification is always a complex piece of work in the case of the higher animals which differ from each other in important features, and between which a comparison can be instituted, the difficulty attending the classification of simple organisms which present only the slightest differentiations is still more difficult.

The principal division made is that which divides them into animal Micro-organisms or Protozoans and vegetable Micro-organisms or Microphytes.

The line of demarcation between these two kingdoms is far from being well defined; there are a great number of micro-organisms *incertæ sedis*, which botanists usually place in the vegetable kingdom, but which zoölogists prefer to classify as belonging to the animal kingdom.*

We give below a list of the most important groups of animal micro-organisms.

ANIMAL MICRO-ORGANISMS.

INFUSORIA.	MASTIGOPHORES.	SARCODINES.	SPOROZOA.
Ciliates	Flagellates.	Rhizopods.	Gregarinida.
Suctoria (Suckers)	Choanoflagellates.	Heliozoa.	Coccidia.
....	Dinoflagellates.	Radiolarians.	Sarcosporidia.
....	Cystoflagellates.	....	Myxosporidia.
....			Microsporidia.

We propose, now, to study the psychic life of these lower organisms, or, to speak in more general terms, their life of relation. It is well known that the expression, the life of relation, comprehends essentially two distinct ideas: first, the action of the external world felt by the organism: or sensibility; secondly, the reaction of the organism on the external world: or move-

---

* The best mark to distinguish the two kingdoms is the chemical nature of the enveloping membrane: in the case of vegetable organisms, the enveloping membrane is made up of a ternary substance, cellulose; while in animal organisms it is albuminoid in character.

ment. It is customary to apply to the union of these two properties the name irritability, which expresses the reaction of the micro-organism upon exterior forces. It is therefore held, and with reason, that every living cell is irritable, that is to say that it possesses the property of responding by movements to the excitations which it suffers.

In admitting then that irritability is the foundation of the life of relation, and consequently also the foundation of psychology, we must nevertheless guard against comparing the autonomous cell of micro-organisms to a simple irritable cell. Although the body of these small beings may be equivalent to a simple cell, it would be an error to believe that their life of relation consists in a motory reaction consequent upon exterior irritation. At the close of our investigations into the psychology of Proto-organisms we shall see that, in these inferior beings which represent the simplest forms of life, we find manifestations of an intelligence which greatly transcends the phenomena of cellular irritability. Thus, even on the very lowest rounds of the ladder of life, psychic manifestations are very much more complex than is usually believed, and the conception of cellular psychology which some very recent authors have formed, seems to me a very crude analysis of the most delicate of phenomena.

In the great majority of pluricellular animals, the life of relation is exhibited in a nervous system and in a muscular system. In Micro-organisms the same cannot be said to be the case: the greater part possess neither a central nervous system nor organs of sense; some even lack organs of locomotion. The functions of the life of relation are performed by the entire mass of the body: many of the Protista, for example, have not a trace of an anatomically differentiated visual organ; it is the entire protoplasm of the elementary organism that is excitable by light, as it is also by heat or by electricity. In other Micro-organisms somewhat higher in the scale, a beginning of differentiation may be seen to make its appearance, giving birth either to some organ of sense or to some organ of locomotion.

We shall give a general description of these organs. The study of this first move in the work of differentiation is of great interest to comparative anatomy and physiology; no less interesting is it to psychology. Besides dwelling on these preliminaries of our work, we shall have occasion to note new and interesting facts.

I.

·THE MOTORY ORGANS AND THE ORGANS OF SENSE.

Motility. From the schedule of the groups of animal micro-organisms which we have given, it will be seen that they are subdivided into four classes, the Infusoria, the Mastigophores, the Sarcodines and the Sporozoa. The distinction between these classes depends on the existence and the nature of the motor organs.

The Infusoria comprise the protozoa that move by the aid of vibratile cilia distributed in greater or less number over their body.

The second class, the Mastigophores, comprises those animals which move by the aid of flagella, that is to say by the help of long filaments.

The third class, the Sarcodines, comprises those animals which move by the aid of pseudopodia; which are projections of the substance of their bodies.

The fourth class, the Sporozoa, is characterized by

the mode of multiplication: they are reproduced by spores. In the animals of this group, the special motor organs are wanting; these creatures therefore generally move very little, or they present only movements of which the principles are unknown.

We shall successively describe the pseudopodia, the vibratile cilia and the flagellum.

The Pseudopod. The formation of pseudopodia takes place chiefly in naked cells—in cells lacking an enveloping membrane, in the Sarcodines in general. They can easily be studied in the *Amœba princeps*, a microscopic animal which is found in abundance in fresh water containing organic matter in a state of putrefaction. It has the aspect of a small gelatinous mass, irregular, formed of a colorless substance, the protoplasm. The chemical nature of protoplasm is still very imperfectly understood; it is only known that it is the result of a mixture of albuminoid matters, with an addition of water and mineral elements. In the protoplasm of the amœba exists a small rounded and refracting mass, containing one or two bright corpuscles in its interior; this small mass is called the nucleus, and the corpuscles the nucleoli.

The form of the body of the amœba is rendered very irregular by the fact that certain parts of the mass lengthen, and form short and rounded protuberances which are designated by the name of pseudopodia. It is by means of these pseudopodia that the animal moves; it emits them in the direction in which it is going, then it retracts them, while other parts of the mass are in their turn elongated. The whole body moves by creeping. The amœba in moving has the aspect of a drop of oil moving along. To explain the mechanism of this movement, it must be supposed

that the extended pseudopod seizes some point of support with its free end, then, in contracting, draws the
entire mass of the body up to this.   But it is difficult
to understand what the cause of the elongation of the
pseudopodia is.   It has been supposed that the pro-
toplasm·is endowed with great elasticity and that the
elongation is the return of this substance to its primitive form.   That is not the explanation given by M.
Rouget.   The learned professor of the Museum has
been kind enough to write out the following note for
us, in which he recapitulates his opinion:

"Every time that a protoplasmic organism dies, or
is subjected either to a strong electric excitation, or
to a relatively high temperature (+ 45° to + 50°), the
pseudopodia are retracted and re-enter into the mass,
which assumes a globular form; the same is the case
in the protoplasm of vegetable cells, the inter-cellular
reticulum of which breaks in receding, or else the mass
of protoplasm divides into spherical bodies.   These
states of retraction are the analogues of *muscular
rigidity,* and like it represent the condition of *maximum*
contraction in the protoplasm—nevertheless the style
of the Vorticels (*Carchesium*) which is a protoplasmic formation, under the same conditions, remains in
a state of permanent retraction.   It follows from this
that the emission of the pseudopodia, *their elongation,*
cannot in any case be considered as a direct act of the
contractility of the protoplasm.

"The production of the pseudopodia, one of the
most difficult problems, cannot, in my opinion, be explained, except in the following manner: All protoplasmic masses, and especially the amœba, consist of
two parts, an enveloping membrane or ectosarc, vis-

cous and elastic, and the central liquid contents holding granules in suspension.

"From the time of the apparition of a pseudopod, a current of liquid is visible which penetrates into the pseudopod and which seems to contribute to its elongation. It is very evident that the liquid is passive, that it penetrates into the pseudopod only because, pressed upon from all sides, it finds less resistance there. I think that the (in appearance) homogeneous hyaline substance of the pseudopod is also a species of *hernia* of the estosarc, resulting from a diminution of the elastic resistance at the point where it appears, with an increase of elasticity or of contractility (to me two modalities of the same property) in those parts of the ectosarc where pseudopodia are not produced. When the contractility or the elastic tension of these parts diminishes, and returns to its original state the pseudopod re-enters into the mass. Add to this that, in an amœba of large dimensions, *Amœba terricola*, it has seemed to me that the most external membrane of the ectosarc showed striæ of a granular appearance which may be identical with the striæ or contractile fibrils of the ectosarc of the ciliated infusoria, *Stentor, Spirostomes, Bursaria*, etc." (May 20, 1887.)

The pseudopod does not represent a permanent, differentiated organ of locomotion; it is produced by a simple prolongation of the mass of the body, which can take place at any point whatever, and when the act of locomotion has been accomplished, this prolongation re-enters into the common mass without leaving any traces of its emission. In other animal species, for example the *Petalobus* of Lachmann, initial traces of differentiation of the pseudopodia have been observed; they always form at the same point of the body, on a level with the anterior part; but, in spite of this constant localization, the motor organ has only a transitory existence; it is produced at the moment it is needed, and disappears into the mass of the body, when the movement has been executed. In the *Actinophrys* there is a still greater progress: the numerous pseudopodia emitted by this animal, and which have the form of filaments, are permanent organs with definite functions.

The Vibratile Cilia. The vibratile cilia are short, extremely thin, homogeneous filaments which are agitated by a vibratory movement. These are distinctly differentiated organs of· locomotion. They have, moreover, several functions: firstly, they enable the animal to move about in the liquid; secondly, they serve it as an organ of prehension; thirdly, they permit a renewal of the water which furnishes the necessary air for respiration to the animal; perhaps they also serve as organs of touch.

The vibratile cilia lend to the Infusoria their peculiar character and enable them to be distinguished from all the other Protozoa. Cilia are also found in vegetable species when young, and in the larvæ of Coelenterates, of mollusks and of worms. But among the Protozoa, it is the Infusoria alone that are ciliated. The cilia are distributed in various manners, differing according to the species. In the *holotricha*, they are distributed regularly over the whole surface of the body, and almost all have the same length; in the *Heterotricha*, they also cover the whole surface of the body, but they are unequal in length. To this group belong the *Stentors* which have long cilia inserted around a circular surface, extending almost to the mouth. This surface is a rotatory organ, analogous to that of the rotifers; it produces eddies in the water and thus causes the flow of foreign bodies to the mouth: these animals have the rest of their bodies covered with fine cilia. In the *Hypotricha* the cilia are located on the ventral surface of the body and aid in locomotion. In the *Peritricha*, they form a circular or spiral row on the anterior part of the body, and lead to the mouth. This is observed in the Vorticels, sessile species which have no other cilia than those which are used for the prehension of food; the rest of the body is bare.

Much has been said about the morphological significance of vibratile cilia; several micrographists have held that the cilia are attached to the enveloping membrane only, and have no connection whatever with the protoplasm. That was notably the opinion of Robin; it is entirely wrong. The cilia are never simple prolongations of the cuticle; they have their root in the protoplasmic substance; they pass through orifices in the cuticle, which consequently is pierced by a multitude of small holes. Engelmann, in recent observations, has been able to trace the extremity of the vibratile cilia into the interior of the protoplasm; he made this observation on the marginal cilia of the Stylonichia; from each of these threads he has seen separate a pale fibre, which moves along almost directly beneath the cuticle in a direction perpendicular to the lateral edge of the body; towards the median line of the ventral face the fibres are often laid bare, because the body of this Infusory voids its protoplasmic substance; there the fibres have the aspect of tightened threads. Engelmann sees in this observation a confirmation of the opinion that the bodies of infusoria are formed of one single cell, because, according to

other observers, there exist also in vibratile cellules
filiform striæ which seems to be a continuation of the
cilia, and which traverse the protoplasm of the cell
throughout its whole length.

We might-add to this direct observation several
other facts showing that the vibratile cilia are indeed
prolongations of the plasm. Under the action of
re-agents the cilia act like the cellular protoplasm;
they are coagulated by the acids and dissolved by
weak alkalies, while the cuticle offers a greater resistance to these same agents.

These vibratile appendices are not without analogy
with the pseudopodia of naked cells; Dujardin, a
French naturalist, demonstrated this in 1835, although
efforts have since been made to bestow the honor of
this discovery upon the Germans. Dujardin has
proved that the amœboid movement and the ciliary
movement are only two manifestations of the contractile power of protoplasm. In fact, if instead
of examining a pseudopod with lobed outline like that
of the amœba, we observe the slender and filamentous
pseudopodia of the Foramenifera, we see that the extremity of the filament is agitated by the same vibratory movement as the vibratile cilium.

All the transitions from the fine and delicate cilia
to the large cilia, tapering in form like a stilleto, which
have been called cirri, have been observed; moreover
these cirri are formed of agglutinated cilia; by the aid
of certain re-agents they have been dissociated.

An observation of a ciliated infusory, the *Didinium
nasutum* (see the illustration further on) made by M.
Balbiani, shows that the movement of the cirri is not
an involuntary movement like that of the cilia of the
vibratile epithelium, with which it has often been

compared, but that it is completely under the control of the will of the animal, like the organs of locomotion of animals much higher in point of organization.

"The *Didinium* has two rows of equal, and rather strong, vibratile cilia, disposed transversely around the body, in the form of two belts or crowns. The rest of the body of this animal is entirely stripped of cilia, but its double vibratory belt suffices to enable it to execute the most rapid and most varied evolutions in the water. Not only does it swim forwards and backwards with perfect ease, but the progression in both directions is always accompanied by a rapid rotatory movement of the animal about its longitudinal axis, similar to that observed in other infusoria that have a cylindrical body. The two rows of cilia always act in union during the locomotion, and the direction which the animal gives to them, determines the direction in which it wishes to move. In the movement for-

Fig. 1.—*Didinium nasutum* (Balbiani) Figure representing movement forward. The cilia are all turned towards the front part of the body.

Fig. 2.—*Didinium nasutum* (Balbiani). Outline of movement backwards. The cilia are all turned towards the back part of the body.

Fig. 3.—*Didinium nasutum* (Balbiani). A sketch of rotatory movement in one spot. The cilia of the anterior belt are directed forwards, while those of the posterior belt are directed backwards

wards, all the cilia are directed toward the anterior part of the body (fig. 1); when it swims backwards, they are reversed (fig. 2). The infusory thus rapidly makes its way across the field of vision by jerks; from time to time it suddenly stops, all the time continuing to turn around rapidly on its

axis on the one spot, during which movement the ciliated belts beat the water in opposite directions, the
anterior ones being turned forwards, while the posterior
are turned backwards (fig. 3). The result of this is
that the effects of these small locomotive apparatuses
neutralize each other in the same manner as two heli-
ces acting in opposite directions, and that the animal
remains stationary, while all the time turning rapidly
about itself, sometimes horizontally, sometimes verti-
cally on its conical appendage, just as on a pivot."

Certain Infusoria, for example the *Condylostoma
patens*, which has been thoroughly studied by M.
Maupas, possess at the same time the two kinds of
appendages, the cilia and the cirri. The former, which
cover the dorsal surface of the animal, are fine, very
dense and animated by a rapid and unceasing vibratile movement. The cirri, which cover the ventral
surface are placed apart; furthermore they do not vi-
brate rapidly; their movements are slow, and when
the infusory moves, one can see them move successively on the plate of glass and support themselves
there, in the manner of a foot, to make the body advance. When the animal stands still, the cirri are absolutely immobile, while the cilia continue their vibratile movement. This observation which can equally
well be made of the Oxytrichid, shows that the vibratile cilia are the organs of involuntary movement, and
that the cirri are more directly subject to the will.
The fact is demonstrated by the experiments of Rossbach, who observed that, under the influence of the
falling of the temperature (from + 15 to + 4) or of
the rising of the temperature (from + 35 to + 40) or
under the influence of various chemical substances,
the large cilia, the organs of voluntary movement, are

paralysed, while the fine and delicate cilia continue their movements, which do not seem to be under the influence of the will. These movements alone cause the whole body to rotate until the vibratile cilia are in their turn paralyzed.

Besides the cilia and the cirri, other appendages in the form of membranes are found among the Infusoria, appendages which are attached to the anterior part of the body or the peristome; these membranes serve the purpose of causing eddies in the water, which bring the floating alimentary particles into the mouth. They are modifications of the vibratile cilia; these membranes like the cirri are formed of agglutinated cilia.

The Flagellum. The study of the third organ of locomotion, the flagellum, brings us to speak of the class of Mastigophores and more particularly of the group Flagellata. The Flagellates are Protozoa of very small size, all in all, very much smaller than the ciliated Infusoria. They have no vibratile cilia at all, but they are always equipped with one or more filamentous appendages which have the form of a long lash. This is the flagellum. This lash, like all the organs of locomotion hitherto studied, has two functions: it is at once an organ of locomotion and an organ of prehension. The flagellum is most frequently single or double (see fig. 4, representing the *Euglenadeses* with its single flagellum); sometimes a person can count a much larger number of them, four, six, eight, ten, and more. As regards the insertion, the same variations are met with. Sometimes the flagella are very numerous and seem to be planted on the same point of the surface of the body, thus forming a brush or plume. In other species we find several flagella arising in the anterior extremity of the body, directed forwards, and also posterior or caudal filaments which are turned toward the rear. This is observed in the genus *Trichomonas;* the anterior flagella serve for purposes of locomotion, perhaps also for the prehension of food; the posterior flagella, on the contrary, are solely organs of locomotion; they resemble a trailing tail and perform the functions of a rudder.

In passing we may point out the great morphological resemblance between the Flagellata and the spermatozoa of animals, the antherozoa and the zoöspores of plants. The organs of propulsion in these beings are the same.

The Protozoan with its flagellum executes the most varied movements, moving first in one direction, then in another, and in different planes; sometimes the animal curves about entirely; but most frequently, when he uses it as an organ of prehension, he extends it its whole length before himself; the basilar part remains completely immovable and rigid, while the free end alone executes movements destined to drive food to the mouth, which is generally situated at the base of the flagellum. Ehrenberg gives to the flagellum the name proboscis; its peculiar mobility renders it worthy of this name. The flagellum, like the vibratile cilium, is an expansion of the protoplasm through the enveloping membrane. M. Certes has observed a Proto-

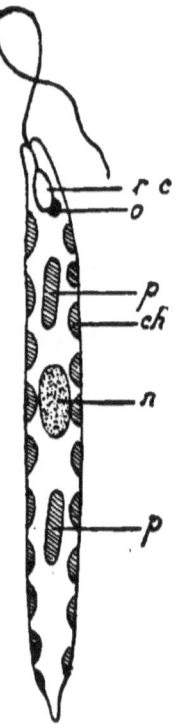

Fig. 4.
Euglenadeses.

r. c. — contractile reservoir; *o.* — eye; *p.* — disk of the paramylone; *ch.* — chromatophores; *n.* — nucleus.

zoan, the flagellum of which between whiles re-entered into the mass of the body, with which it mingled; it was replaced by a pseudopod which soon attenuated and took the form of a flagellum.

Bütschli has recently made a very interesting observation on this organ of locomotion. Under certain circumstances, the Peridinia (Dinoflagellates) throw off their long flagellum and enter into a state of repose; they generate them quite as easily. In the *Glenodinium cinctum*, Bütschli has seen the flagellum roll itself up first like a cork-screw, and then suddenly detach itself from the animal; having become free, it stirs about in the water for several minutes before becoming motionless. This observation enables us to refute those naturalists who believe that the vibratile cilium is an appendage of the cuticle, by bringing forward the fact that when the cilia with the portion of the cuticle in which they are inserted are separated from the cell, the cilia continue to move; we have just seen that the flagellum moves even after it is separated from the cuticle; this persistence of movement is sufficiently explained by the protoplasmic nature of the cilia and of the flagellum.

From another point of view, the observation of Bütschli gives us a curious example of the phenomena of *autotomy*, which have recently been studied by Frédéricq.

The pseudopodia, the vibratile cilia, and the flagellum, constitute the three motor organs that are most frequently found in the kingdom of the Protista.' Among the Infusoria, moreover, particular differentiations of the protoplasm have been described, which may be compared to the muscular fibres of the higher animals. The Vorticellæ are supported by contractile peduncles. These are filaments capable of rolling themselves up into the form of a cork-screw, when the animal is disturbed. Certain Infusoria can modify the form of their body by a sudden contraction: they have been called *metabolic*; such are the Stentors, the Prorodons, the Spirostomes. In contradistinction, those which do not change their form, for example the Paramecia, have been called ametabolic. According to the observations of Lieberkühn, which date back to 1857, the metabolic Infusoria have their bodies divided into large granulous bands, separated by bright filaments. It has been asked which is the contractile element: is it the band, or is it the filament? Oscar Schmidt, Kölliker, Stein, and Rouget think that it is the band which is the contractile element. This opinion is based on the following fact, which M. Rouget was the first to observe: at the moment at which the animal contracts, the band presents transverse striæ; this appearance is due to the fact that the bands contain in the state of rest small granules which, during the contraction of the animal, are disposed in transverse series, so as to recall the *sarcous elements* of Bowman.

Lieberkühn, Greef, and Engelmann attribute the active part to the bright fibre. Engelmann has based his opinion on the fact that he recognized in the filament the property of double refraction, which, according to him, belongs to all contractile substances, while the substance which separates the filaments shows only single refraction.

However that may be, it is one of these two elements that possesses the power of contraction, and which deserves the name of *myophane*, which Haeckel gave it. It is very remarkable that in the Stentors and the Spirostomes the fibrillous striæ are in intimate connection with the basilar extremity of the vibratile cilia. In the Vorticellæ one can clearly see the fibrils converge toward the axis of the style, the contractile element of which they constitute.

We shall not leave the study of the motor organs without saying a word about the rhythmical movements which can be seen in the contractile vesicle of the Micro-organisms, vegetable as well as animal. This vesicle is a small cavity which is dug into the proto-plasm, and which alternately increases and diminishes its capacity. Scientists by no means agree as to its ex-act function; Bütschli and Stein consider it to be a secretive apparatus. Its pulsations are very regular. Their number is constant in every species. In the *chilodon cucullulus*, a pulsation occurs every two sec-onds; in the *Crytochium nigricans*, every three sec-onds; in the Vorticellæ, every eight seconds; in the *Euplotes*, every twenty-eight seconds; in the *Acineria incurvata*, every six minutes; Rossbach, whose curi-ous experiments with the vibratile cilia and the cirri we have already cited, has made analogous experi-ments with the contractile vesicles. He observed es-pecially that, under the action of alkaloids, the con-tractile vesicle ceased pulsating in diastole, and di-lated enormously; but poisonous agents do not act all at once on the movements of the vesicle; they begin by paralyzing the larger cilia, which are under the in-fluence of the will. The movements of the vesicle, like those of the small cilia, persist for a much longer time. M. E. Maupas has seen Paramecia, killed by a discharge of trichocysts, become completely immo-bile, with their vibratile cilia inert and rigid, while the contractile vesicle continued to pulsate with the same activity; this activity continued for an hour.

We have now briefly examined the morphology of the motor organs of Micro-organisms.

It is very difficult to determine the physiological process of the movements produced by these organs. The simplest movements and the ones most easily understood, are those by which a cell suddenly and strongly irritated withdraws its prolongations and assumes a spherical form; this change of form can be explained by a quick condensation of the protoplasm, which becomes the seat of a phenomenon similar to that of a contracting muscle. The sudden modifications which are observed to take place in the form of the so-called metabolic Infusoria are in this way explained by an analogous phenomenon, so much the more evident as the Infusoria which possess this property, show in the cortical layer of their protoplasm (ectosarc) granulous bands which have with more or less justice been compared to the muscles of the higher animals. The displacements of the body determined by the pseudopodia, by the vibratile cilia, and by the flagellum are much more difficult to interpret; meanwhile it is probable that the movement proceeds from the contractions of the protoplasm which are produced either in the ectosarc or in the motor organ itself; the latter is automobile, as is seen, for example, when a flagellum separated from the rest of the body continues to move in the liquid.

It is well known that any number of discussions have been raised as to the manner in which the pedicel on which the Vorticellæ are mounted, contracts. Still more obscure is the oscillatory movement of the Bacteria. These small beings are very mobile when they find themselves in·a liquid; they frequently exhibit a movement of oscillation which sometimes carries them forward, sometimes backwards. An attempt has been made to explain these movements by postulating the presence of organs of locomotion, extremely slender filaments planted at one of the extremities of the Bacteria like small rods; but the existence of these organs has not been absolutely proved. Even more obscure is the movement observed in certain Gregarines.· It would seem that in the case of these animals, which are often of considerable size, one ought to be able to understand the principle of their movements much more easily than in the case·of such small beings as the Bacteria; but this is not the case. The Polycystids have a very peculiar manner of moving; the motion is one of perfect translation, uniform and rectilinear; the animal seems to slide all of a piece over the object-plate; it can go to the right, to the left, stay its motion and resume it again; it is. free in directing its movements. Now, during this movement nothing can be seen to take place in the body from within or without. An analogous phenomenon is to be observed in the Diatomes. Some scientists have wished to explain the mysterious motion by translation executed by the Gregarines, as being due to an imperceptible undulation of the sarcode; but if there were any undulations whatever, one ought to observe a correlative movement in the granules inside; now this is something that is never seen.

Thus there still exists a great deal of obscurity concerning the principles determining motion among the Proto-organisms. The theories based upon muscular contraction that have been propounded from observing higher animals, are by no means sufficient to explain the phenomena of motility among certain Protozoa and Protophytes.

Nervous System. Hitherto not the minutest trace of a central nervous system has been found in a single Proto-organism. The nervous function among these inferior beings devolves upon the protoplasm, which is irritable, which feels and which moves, and which, in certain species, as we shall see later on, is even capable of performing certain psychic acts, the complexity of which seems quite out of proportion to the small quantity of ponderable matter which serves as a substratum to these phenomena. There is, moreover, no occasion to be surprised that an undifferentiated mass of protoplasm should be able to exercise the functions of a veritable nervous system. In fact every nervous element is nothing else than the product of protoplasmic differentiation; the protoplasm embodies in itself all the functions that, in consequence of an ulterior division of labor among the pluricellular organisms, have been assigned to distinct elements.

It has rightly been held, therefore, that if no nervous system, anatomically differentiated, existed in proto-organisms, it must be admitted that their protoplasm contains a *diffused nervous system.* Among all the observations that uphold this idea, we must cite one to which M. Gruber, a professor at Freiburg, in Breisgau, has recently called attention. This observation was made on a large, ciliated Infusory, the Stentor, of which mention will be made so often hereafter that it will be advantageous to give a full description of it beforehand.

The Stentor has an elongated body, broadened in front like a funnel, and able to fasten itself by its posterior extremity. The edge of its peristome is covered by a belt of vibratile cilia disposed about a spiral line. The mouth occupies the most sunken part of the peristome.

The body of the animal is striated with longitudinal bands; at the plane of the peristome, these bands take a different direction: they become transversal and spiral. In the interior of the protoplasm can be observed a contractile vacuole and a nucleus like a string of beads, made up of a large number of grains. This

Infusory, like all the Ciliates, multiplies by fission; a contraction is seen to take place in the middle of the body; the segment below the contraction generates a peristome similar to that of the upper segment; then a second contractile vacuole is formed, and soon the two segments represent two complete animals which possess all their organs. Nevertheless, the two Stentors continue to be united for a certain length of time by a bridge of matter, located even with the point where the contraction took place; this bridge of matter gradually grows thinner and thinner and becomes as fine as a thread. (See fig. 5.) Now, Gruber has observed that the two Stentors united by this bridge of protoplasm exhibit perfect

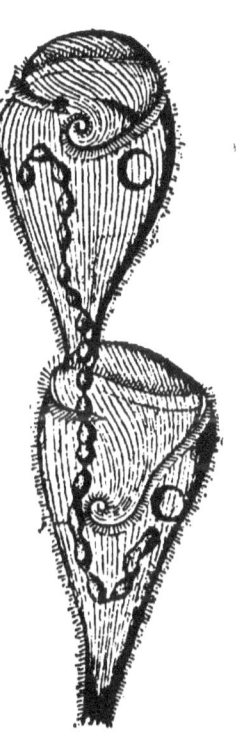

Fig. 5. Stentor in process of division.

harmony in their movements; they always sway in the same direction at the same time; and this harmony is necessary, because the least contrariety of motion would suffice to break the feeble bond that unites them. Moreover, their vibratile cilia beat in unison. To explain this concordance in the movements of the two animals, Gruber assumes that the entire mass of their protoplasm performs the function of a diffused nervous system, which has the effect of regulating their movements and of making them harmonize.

We might add that the Infusoria possess not only a diffused nervous system, but that they must of necessity possess special nerve centres, endowed with different functions.

It will be remembered in fact that, under the influence of certain poisonous agents, death is not simultaneous throughout all parts of the organism. What ceases first are the voluntary movements of the large cilia; the movements of the small cilia are able to persist much longer; and finally, when all the cilia have become immobile and rigid, the vesicle has still been seen to pulsate for an hour. This gradual death recalls what we remark among the Vertebrates; under the influence of poisonous agents, the brain dies first, then follows the marrow, and lastly the bulb, which is the *ultimum moriens*.

The Organs of Sense. All the Micro-organisms are endowed with sensibility; some, like the Infusoria, have exceedingly sensitive powers. But, hitherto, organs of sense anatomically differentiated have been found in only a very small number of species. Generally, the protoplasmic expansions which we have above described under the name of pseudopodia are regarded as fulfilling the function of rudimentary organs of touch which advise the micro-organism of the presence of objects which happen in its path; but these pseudopodia, which at the same time serve as motor apparatus, do not exhibit any structure which especially fits them for the reception of sensory impressions. Similarly, Stein considers the vibratile cilia as organs of touch. As these are organs which have not undergone any differentiation, we shall not stop to consider them. The Infusoria belonging to the genus *Cryptochilum* (Maupas) carry at their posterior extremity a long rigid bristle, which M. Maupas regards as an organ of touch, intended to advise the animal of the approach of other Infusoria.

We shall speak more at length of the organ of sight; this has been the subject of numerous treatises, some of which are quite recent and of the greatest interest to general physiology and psychology. Of all the organs of sense the eye is the one which is first differentiated. It is found in the organisms belonging to the vegetable kingdom as well as in those belonging to the animal kingdom. While these small beings do not seem to possess any organ especially adapted by its structure for the reception of tactile, olfactory, or gustatory impressions, a large number already exhibit an ocular spot, that is to say a differentiated organ, for the purpose of sight and for no other purpose.

Let us first turn our attention to the eye of the Protozoa.

It is chiefly in the group of Flagellates, and principally in the species that are colored green by chlorophyl (for example the Euglenæ), that ocular spots are found; these spots which are colored a bright red, present themselves very clearly to the observation, for they are set off by the uncolored plasma of the anterior part of the body where they are generally located. Oculiform spots are also found in the species

colored by yellow chlorophyl (*Uroglena volvox*, etc.).
Generally, there is only one spot, situated at the base
of the flagellum. This is seen especially in the *Euglena
viridis*, a small flagellate infusory, which is very
abundant in fresh waters, which it often covers with
a thick green coating.

In the *Synura uvella*, a colony-forming flagellate,
there exist in each individual, in the anterior part of
the body, numerous spots, varying from two to ten.

Below we give an illustration representing the
anterior extremity of the *Euglena Ehrenbergii*, according to Klebs. A large ocular spot is noticeable,
in contiguity with the contractile reservoir. Ehrenberg, deceived by the appearance of these two organs, had taken the contractile reservoir for a nerve
ganglion.

It is not only in the large
group of Protozoans that the red
spots are met with; they are
found also among the vegetable
Micro-organisms. A large number of green-colored zoöspores
exhibit at the anterior, and
usually colorless, extremity of
their bodies, a small red point
which seems to have exactly the

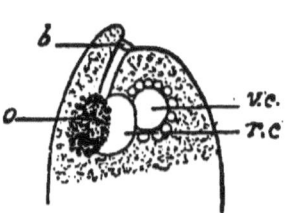

Fig. 6.—Anterior extremity of
the *EuglenaEhrenbergii* (after
Klebs). *b.* ━ Mouth and gullet.
━ *o.* ━ Eye. ━ *v. c.* ━ Contractile vesicle. *r. c.* ━ Contractile reservoir.

same structure as the red spot of the Euglenæ. It was
on this fact that Stein based his opinion that the
spot of Euglena is not an eye; to him it seemed impossible to admit that the vegetable Proto-organisms
could possess a visual organ. This is an excellent
instance of *a priori* reasoning. Later on we shall
see that Stein's view has now been completely
abandoned; the very opposite view is taken, for the

eye of the Protista is considered as being destined to perform chiefly a vegetable function.

Klebs was able to study the structure of the ocular spots, by employing a very ingenious artifice. When the Euglenæ are treated with a solution of sea salt, in the proportion of one part to one hundred, an enormous dilatation of the contractile vesicle, which forms a hollow in the protoplasm of the animal, is induced; now, as the red spot is, so to speak, glued to the vesicle, it undergoes the same dilatation as the latter does, thus greatly facilitating observation. By this treatment it has been observed that the spot is a small discoid or triangular mass, of jagged and irregular outline; it is formed of two material parts; for a base it has a small mass of reticulated protoplasm, and in the meshes of the protoplasm there are small drops of an oily substance, colored red.

This red pigment, which has received the name of hematochrome, is not without its analogy with the green pigment of the chlorophyl, because this latter becomes red under certain conditions. For example, the chlorophyl pigment which fills the entire body of the *Hematococcus pluvialis* becomes red, when the animal enters into a state of rest; the stagnant spores of the algæ also assume a red tint. So, also, in numerous plants, the parts of the flower destined to become red are green as long as they are enclosed in the bud. It is thus probable that the red pigment of the Euglenoids is derived from a green pigment.

What is the physiological significance of these spots? Ehrenberg considered them as eyes; hence the name *Euglena* (word for word, pretty eye), which he had given to a species of Flagellates provided with ocular spots. This interpretation had been questioned by all the authors of his time, especially by Dujardin. At the present day, however, naturalists have come back to it, in consequence of observations which have been made on other Micro-organisms that possess a more perfectly developed eye.

M. Pouchet has discovered in the *Glenodinium polyphemus*, which belongs to the group of Peridinia (or Dinoflagellates, according to the classification of Bütschli), an eye about the function of which there can be no mistake.

This eye occupies a fixed place in the cellule of the Peridinium; it has a uniform location and position. It consists of two parts, the one a veritable crystalline humor, and the other a veritable choroid. The crystalline is a strongly refracting, hyalin, club-shaped body, rounded at its free end, which is always directed forwards, while the other end is immersed in the mass of pigment which represents the choroid. This latter is clearly determined; it forms a sort of hemispherical cap, enveloping the posterior extremity of the crystalline. In one of the two forms of *Glenodinium polyphemus*, the choroid pigment is red; in the other it is black.

M. Pouchet has been able to establish that in the young animals the crystalline is first formed of six to eight refracting globes, which are merged into each other in order finally to constitute one unified mass. Also, the choroid is the result of a combination of the pigmentary granules which, at first sparse, group together and finally form the hemispheric cap that covers the posterior extremity of the crystalline.

In fact, the visual organ of this Peridinium is composed of exactly the same parts as the eye of a metazoön with one exception, the absence of the nerve element. This is not at all differentiated, but remains diffused, like the whole nervous system. M. Pouchet calls attention to the interest which his observation affords from a taxonomic point of view. The Peridinia have sometimes been classed among the vegetables; the presence of starch and of cellulose in their protoplasm has induced Warming to classify them among the Diatomaceæ and Desmidiaceæ. It is admitted to-day that certain Peridinia possess an eye, an organ which has hitherto been considered as the exclusive attribute of animals. Nothing more clearly emphasizes the altogether artificial character of the distinction between animals and vegetables than the results of dealing with Micro-organisms.

Before leaving the Peridinia, we would remark that these small beings afford an interesting fact from the point of view of the history of the Protozoa; they are provided with a long flagellum; they exhibit in addition an equatorial line on which formerly a crown of vibratile cilia was thought to be recognizable: this supposed co-existence of a flagellum and of cilia had determined the naturalists to form a group of Cilio-flagellates, serving as a transition between the Flagellates, properly so-called, and the Ciliates. Since then it has been discovered that the Peridinia do not possess vibratile cilia; what had given rise to this error is the presence of a second flagellum on the level of the transverse line which we have just described; the movements of this flagellum have the appearance of vibratile cilia in motion.

Some time before the investigations of M. Pouchet, M. Künstler (of Bordeaux) had discovered, in a Flagellate of the genus *Phacus*, a red eye which is also formed of two parts; it is composed of a homogenous globule, acting as a crystalline humor, and surrounded by a red pigment, acting the part of the choroid.

Before M. Künstler, Claparède and Lachmann, in their important work on Infusoria and Rhizopods, had described a similar visual organ in the *Freia elegans*, a ciliated infusory of the family of Stentorines. "Immediately behind the point of truncation," say they, " there is found a lunate spot of intense black, evidently belonging to the category·of those phenomena which M. Ehrenberg, in the Ophryoglenæ, for example, calls an eye or an ocular spot. The significance of this spot has never been known. It was often very much denser than that of the Ophryoglenæ, and sometimes there was discovered behind it a very transparent corpuscle, which involuntarily gave rise in the mind to the idea of a crystalline humor. We cannot, however, add much of importance to this idea, since the functions of a refracting apparatus must necessarily remain problematic, as long as we do not discover behind it a nervous apparatus fitted to perceive the impressions received."

This last conclusion seems to us excessively cautious. The co-existence of a pigment and of a crystalline humor amply suffices to characterize a visual organ. As to the nerve apparatus susceptible of perceiving impressions, it is replaced by the protoplasm, which, as is well known, is sensitive to light.

Even before that, in 1856, Lieberkühn had discovered in a ciliated infusory, the *Panophrys flavicans*, an ocular spot, composed of a convex crystalline humor, having the form of a watch-crystal enveloped by pigment and placed on the convex side of the oral fosse. In another species, the *Ophryoglena atra*, he found black pigment, but no crystalline humor.

It is impossible to believe that these organs are not eyes, for ·they have the same structure as the eyes of comparatively higher classes of animals, such as certain worms, turbellaria, rotifers, lower-class crustaceans, etc; all these organs are similarly formed of a small crystalline globule enclosed in a small mass of pigmentary matter. The identity of structure naturally leads to the assumption of the identity of functions.

The eye of the Euglena is the simplest of all; it is even reduced to the maximum point of simplicity, as it is composed of a spot of pigment. What induces us to believe that this spot is a visual organ, is the presence of this pigment. In fact this pigment is found in the most elementary visual organs. A second argument might be advanced; the red pigment of the Euglena exhibits the same re-actions as the coloring matter that fills the rods of the retina in the Vertebrates. From among these re-actions common to both, we cite the decoloration under the influence of light (Capranica).

Whatever the case may be, one thing is certain, namely that the Euglena is very sensitive to the light. When they are kept in a vessel, they are invariably seen to cover the side exposed to the light. M. Engelmann has observed that light acts very strongly upon this small animal; it does not act directly on the spot of pigment, nor, as was formerly thought, on the flagellum, but on the protoplasm which is located in front of the spot. The special microspectral object-glass that M. Engelmann constructed, enables us to see that the Euglenæ always congregate in the band F to G of the spectrum.

So far as the vegetable Micro-organisms are con-

cerned, we have already mentioned that a large num·
ber of the algæ zoöspores exhibit, in the anterior part
of their body, ocular spots of a beautiful ruby color:
these are organs that probably have the same structure as the red spots of the Euglenæ. Moreover, it
is probable that certain Microphytes possess more
complex visual organs, composed of red pigment and
of a crystalline humor. M. Balbiani has recently
testified to this fact·in the case of the *Pandorina mo-
rum*, a spherical colony of green micro-organisms; in
each colony there exists a certain number of individuals which possess a red spot, the shape of which is
perfectly circular; if this spot be examined under a
glass of very high magnifying power, one can readily
see that it is formed of a small spherical globule, covered, on a portion of its surface, by a cap of red
matter. This observation is all the more interesting
because it is made on a being, the vegetable nature of
which is to-day no longer doubted; the Pandorina are
Volvocinæ which modern botanists place among the
algæ. (We are glad to give our readers the earliest
communication concerning this fact.)

In describing the eye of the Protista, we said that
the eye is the only organ of sense which is distinctly
differentiated in these lower beings. But, perhaps,
this assertion is too sweeping. Some species appear
armed with small organs which could easily be in-
vested with a sensory function. In this respect, we
may cite the *Loxodes rostrum*, a beautiful ciliated infusory, remarkable for its proboscis and for the muscular sheath which closes its mouth. This animal
exhibits along the dorsal surface a row of small organs
which, by their structure, seem destined to act a part
in performing the function of hearing. They are

formed of a vesicle, the centre of which is occupied by a refracting globule; they are called the vesicles of Müller, after Johannes Müller, who discovered them. The auditory organs which have been observed in Worms and the Cœlenterata are apparently composed of a vesiculiform capsule enclosing a solid concretion, called otolith. Thus it is possible that the vesicles of Müller may be auditory vesicles. Up to the present time this organ has not been met with in any other species of Protozoa.

II.

NUTRITION.

After studying the organs, let us pass to a study of their functions.

It is not our intention to devote special chapters to irritability, instinct, memory, reasoning, and the powers of volition in Micro-organisms. This would lead to diffuseness of treatment. Our method will be quite different. We shall describe as a whole all the different manifestations of psychical activity attendant upon the actions of Micro-organisms in the exercise of the important functions of their existence. The present chapter will be devoted to psychical phenomena connected with the act of nutrition.

All living matter possesses the power of continually increasing its mass by the inward reception of materials, and of simultaneously decreasing the same through the combustion of its substance with the oxygen of the atmosphere. The first of these processes is called nutrition, and the second, respiration.

We shall first examine the psychical phenomena which precede and determine the act of respiration. These phenomena are often very simple and of little significance. If the Micro-organism lives in the water, which is most frequently the case, the oxygen contained in solution therein passes directly through the cellular cuticle by dialysis and comes in contact with the body of the protoplasm; in which case the process of respiration is solely a chemical phenomenon. But it may happen that a minute organism chances into a medium containing little or no oxygen-gas; amid these new conditions where it becomes necessary to move towards sources emitting oxygen by voluntary effort and directed motion, it has been discovered that a great number of Micro-organisms, and particularly Bacteria, are capable of detecting the expansive power exerted by oxygen in the liquids in which they are found. When bacteria of putrefied matter are put in a drop of water containing no oxygen but in which have been placed chlorophyl algæ, or green Euglenæ, or grains of chlorophyl obtained by crushing green cellules, nothing happens in the first instant; but if the preparation be illuminated so as to allow the chlorophyl to act, the bacteria are seen to exhibit very rapid movements and to proceed, altogether, towards the points of the preparation where the generation of oxygen is taking place, that is to say, about the grains of chlorophyl. Under these conditions a chemical exchange is instituted between the chlorophyl and the aërobious Bacteria: the Bacteria disengage carbonic acid gas and absorb oxygen; the chlorophyl fastens upon the carbon of the acid and sets the oxygen at liberty. If the preparation be darkened the Bacteria cease assembling about the chlorophyl grains, which, hid from the light, cease to disengage oxygen. The clustering begins anew, if a ray of sunlight is again let touch the chlorophyl.

Analogous facts have been observed under circumstances somewhat different. In a preparation from
the intestines of a silk-worm, M. Balbiani has seen
Bacteria which were uniformly distributed throughout
all points of the preparation, gather about the green
and undigested cellules of the leaves contained in the
intestines, and bury themselves in them as if to par-
take of them. In other instances, the same naturalist
has observed that Bacteria developed in a drop of
silk-worm's blood, would gather, after a while, about
the globules of the blood; undoubtedly for the purpose
of seizing the oxygen being absorbed by them.

Upon the basis of these facts M. Engelmann has
established the method called the Bacteria method.
He regards bacteria as a living reagent which enable
us to reveal the trillionth part of a milligram of oxygen, that is to say, a quantity scarcely greater, according to the calculations of physicists, than a molecule.
This curious method enables us to explain biological
problems which had hitherto remained unsolved.
Before this, it was not known whether the colorless
protoplasm of green plants could or could not disengage oxygen. It is now known, thanks to the bacteria, that grains of chlorophyl are the only points about
which the liberation of oxygen takes place. The same
method has enabled us to prove, in the variegated
plants, that the maximum liberation of oxygen coin-
cides with the maximum absorption of light. Thus,
in the case of green algæ, the red and the violet colors
of the spectrum are the spots where the bacteria ac-
cumulate the thickest; consequently here is where the
liberation of oxygen is greatest. Now, these colors
correspond to the lines of greatest absorption in the
spectrum of chlorophyl. In the case of brownish yel-

low cellules, the maximum action is in the green; in the case of bluish green cellules, in the yellow; in the case of red cellules, in the green. The author has concluded from this that there exists a series of coloring substances which, like chlorophyl, have the power of resolving carbonic acid gas; he calls them chromophyls. In the same way, moreover, this method enables us to solve the question of the distribution of energy in the solar spectrum. As M. Engelmann has remarked, it is interesting to see the Bacteria come to confirm our theories as to the composition of solar light.

Bacteria are not the only organisms that eagerly make towards points where oxygen is to be found. A large number of other Micro-organisms act in the same way when they happen into a medium lacking oxygen. M. Ranvier has noticed that if a preparation containing leucocytes, screened from air, be examined for a certain length of time the cellules will be seen to throw out long filaments towards the part that faces the air-side of the preparation. It appears, then, that a rudimentary oxygen-sense exists in the protoplasm of Proto-organisms.

This sense does not merely apprise the organism of the presence of oxygen; it enables it, further, to gauge the tension (expansive power) of the gas. So that, when the tension becomes too powerful, the organisms are seen to flee before it.

III.

The mode of nutrition among Micro-organisms is not uniform—a fact which ought not to appear remarkable when we bear in mind that this immense group is made up of all manner of heterogeneous beings that

have nothing in common save the microscopic little-
ness of their bodies and the simplicity of their
structure. Three main types of nutrition may be
briefly distinguished.

1. *Vegetable nutrition,* or according to Bütschli's
expression, *holophytic.* This is the method of nutrition among animal or vegetable cellules that contain
chlorophyl and that nourish themselves by forming
organic nutriment from ingredients taken from the
surrounding medium. It is hardly necessary to call
to mind that the function of chlorophyl is that of nutrition and not of respiration. This phenomenon was
formerly termed the diurnal respiration of plants. The
expression involves several mistakes. Enough to say
that vegetables respire as animals do, by uniting with
oxygen, and that that respiration continues the same
both day and night. The function of chlorophyl is by
no means respiration; its office is to decompose the
carbonic acid gas of the air and to seize the carbon,
which serves the plant in forming ternary or qua-
ternary substances. This chemical work is performed
by all chlorophyl organisms when influenced by the
radiation of light.

Chlorophyl does not belong exclusively to the vegetable kingdom. A large number of animal Micro-
organisms are colored green by this pigment; they are
met with principally in the important group of Flagellates. Their assimilative organs, which are likewise found in all green plants, bear the name of chromatophores; they have lately formed the subject of
-interesting investigations.

The chromatophores are small bodies of protoplasm
which are distinguished from protoplasm in general
by their having assumed an individual structure.

These little bodies, which in the vegetables are called *leucites*, have a granular and reticulate structure; they are impregnated with a coloring substance, at times green, at times yellow, and àt times brown, as the case may be; in fact, several coloring substances are present, which, by intermixture in different proportions, form colors of many varieties. The best known, after green chlorophyl, is yellow chlorophyl or *diatomin*. The latter coloring substance can be absorbed by alcohol.

The Euglenoididæ, the Chlamydomonadidæ, and the Volvocinæ exhibit enormous chromatophores. In the case of the Euglenæ, the chromatophores are formed of small discoid plates; they are situated directly under the cuticle, so that the light can act upon them (see fig. 4). In certain species of Flagellata, they are exhibited under the cuticle in thé form of two large plates which envelop the protoplasm like a cuirass formed of two pieces. The Chlamydomonadidæ and the Volvocinæ have green chromatophores, disc-shaped, and very small.

In the centre of the chromatophore a small bright space is observed which was formerly thought to be filled with chlorophyl; in reality, it is a minute solid globule which shows an extremely close analogy with the substance composing nuclei, or nuclein. It exhibits the same chemical reactions; it actively absorbs coloring matter and grows extremely brilliant when treated with acids. Schmitz gives this little body the name of pyrenoid (from πυρήν, nucleus). It is around the pyrenoid, and probably through its action, that starch forms; it is deposited in grains or re-unites in a ring about the pyrenoid, a fact easily ascertained by coloring them with iodine.

Production of starch has also been observed in the colorless Flagellates, as for instance in the *Polytoma uvella*. These latter do not have chromatophores, but Künstler, and after him Fisch, has noticed that every grain of starch is attached to a small mass of colorless protoplasm which is the focus of formation for the grains. This is precisely what happens in vegetable organisms where colorless starch-leucites are found. This little mass of protoplasm always faces the hilum of the starch-grain.

As the function of the chromatophores is exercised only when subjected to the influence of light, it follows that green Micro-organisms must have light in order to nourish themselves.

A quite remarkable fact may be adduced in this connection. On examining the kingdom of Protozoans as a whole, it will be seen that a striking coincidence exists between the presence of the eye and the presence of the chlorophyl pigment. Organisms having an ocular spot are in most cases provided with the chlorophyl pigment, or, in other words, nourish themselves as plants do, by generating starch through the action of light. This fact proves that sensibility to light is in some manner dependent upon the chlorophyl function. If Flagellates possessing chromatophores, that is organs generating starch, have ocular spots at the same time, it is because these rudimentary eyes enable them to find their way towards the light, which is the necessary agent of chlorophyl action. Accordingly, all Micro-organisms having eyes nourish themselves as plants do. In their case, the object of the eye is to direct the performance of a vegetable function.

It is interesting to note in this connection that the

Euglenæ might nourish themselves as animals do, for they have a mouth and a digestive apparatus. The buccal, or oral, aperture opens in the anterior end at the base of the flagellum, and is connected with a short gullet or esophagus (see fig. 6, the mouth and gullet of an Euglena). Nevertheless, the Euglena is never seen using its mouth for swallowing alimentary particles. A quite curious problem is involved here. If it is true, as has been claimed, that it is the function that makes the organ, how do we explain the existence and especially the genesis of this digestive apparatus which performs *no* function?

It is the presence of chromatophores that prevents certain Flagellates from feeding like animals; so much so in fact, that the digestive apparatus performs its functions in Flagellates which have no chromatophores and are not provided with chlorophyl pigment, an instance of which is seen in the *Peranema*. The *Peranema* is, further, an exceedingly voracious animal. We must note also that the *Peranema* does not exhibit ocular spots like the green Euglena; and moreover, it has no need of such, since it does not have to seek the light to generate starch. All these phenomena are interdependent.

The influence exerted by light upon the green organisms of both kingdoms has been ascertained by different scientists. Light at a certain degree of intensity attracts them, and at a greater degree, repels them. Some years ago M. Strassburger conducted a series of connected experiments upon the movements of green spores towards light. It was observed, here, that the grains of pigment in the interior of the cellules, when under the influence of solar radiations, executed movements and set outwards in all directions.

2. *Nutrition by endosmosis*, or *saprophytic*. The organism nourishes itself by absorbing through the whole surface of its body liquids containing the products of vegetable or animal decomposition. Saprophytic beings are found in putrid waters or in infusions. This manner of nutrition may be considered, from the point of view which now engages us, as the most simple of all; it probably allows of a search for food, but it is certain that no movements are involved which are designed to draw the food into any possible digestive apparatus.

3. There is now a last mode of nutrition, of which we shall treat in minute detail; namely, *animal nutrition*, where the Micro-organism seizes solid alimentary particles and nourishes itself after the fashion of an animal, whether it be by means of a permanent mouth or by means of an adventitious one, improvised at the moment of need. This manner of nutrition is the process employed by higher animals. Among the lower organisms, it is met with in most of the Infusoria, in the Sarcodines, in many of the Mastigophores, and in others. Respecting the Micro-organisms belonging to the vegetable kingdom, we find nutrition by endosmosis and chlorophyl nutrition; the Protophytes never possess a mouth and never absorb solid foods.

Animal nutrition requires very remarkable psychological faculties in the organism practicing it. These manifestations of psychic life, the progressive complexity of which we intend to trace in starting from the simplest protozoic forms and arriving at the higher—prove that these animalcula are endowed with

memory and volition. We shall group our remarks
under the two following heads:

a. The choice of food; and

b. The movements necessary for the prehension of
food.

The Micro-organisms do not nourish themselves
indiscriminately, nor do they feed blindly upon every
substance that chances in their way. Also, when they
ingest food through some point or other of their bodies,
they understand perfectly how to make a choice of the
particles they wish to absorb. This choice is sometimes quite well defined, for there are species which
feed exclusively upon particular foods. Thus, there
are herbivorous Infusoria and carnivorous Infusoria.
Among the herbivorous ones may be classed the chilo-
dons which feed upon small Algæ, Diatomaceæ, and
Oscillaria. The parmecia live principally upon Bacteria. The Leucophrys is a specimen of the carnivorous class; it devours even the smaller animals of its
own kind. The *Cyrtostomum leucas* eats everything,
as do the Rotifers.

Though the fact of an exercise of choice in taking
food is settled beyond question, yet the interpretation of this phenomenon is a matter of much uncer-
tainty. Some writers, as Charlton Bastian for instance, explain this choice of food as an affinity of
chemical composition existing between the organism
and the nutriment. This idea does not lead to anything. Others compare the discrimination made by
the Proto-organism between objects presented to it,
to the action of a magnet which in some way selects
particles of iron that have been mixed with particles
of other substances. ' The latter interpretation is an
evidence of the tendency evinced by some naturalists,

of endeavoring to identify the attributes of living organic matter with the physico-chemical properties of the mineral kingdom.

In our opinion, the only question demanding consideration is whether the choice of food, in the case of Proto-organisms, does or does not result from a psychical operation, similar, for example, to that which takes place in higher organisms. We have received a noteworthy communication from M. E. Maupas, upon this subject, which tends to establish that the choice of food is not the result of individual taste in the Micro-organisms, but is determined by the organic structure of their buccal apparatus which does not allow them to receive other forms of nutriment.

We must closely examine, therefore, the mechanism for prehension of food.

The following is what occurs when the Amœba, in its rampant course, happens to meet a foreign body. In the first place, if the foreign particle is not a nutritive substance, if it be gravel for instance, the amœba does not ingest it; it thrusts it back with its pseudopodia. This little performance is very significant; for it proves, as we have already said, that this microscopic cellule in some manner or other knows how to choose and distinguish alimentary substances from inert particles of sand. If the foreign substance can serve as nutriment, the Amœba engulfs it by a very simple process. Under the influence of the irritation caused by the foreign particle, the soft and viscous protoplasm of the Amœba projects itself forwards and spreads about the alimentary particle somewhat as an ocean-wave curves and breaks upon the beach; to carry out the simile that so well represents the process, this wave of protoplasm retreats, carrying with it the foreign body which it has encompassed. It is in this manner that the food is enveloped and introduced into the protoplasm; there it is digested and assimilated, disappearing slowly.

There are cellules found in the inner intestinal walls of lower animals which effect the prehension of solid foods in the same manner as the Amœba cellule: they are called phagocytes.

This mode of prehension is beyond contradiction the most simple imaginable; for the prehensile organ is not as yet differentiated. Every part of the protoplasm may be made to serve as a digestive cavity in enveloping the foreign substance.

From the special standpoint of prehension of food, we may place the *Actinophrys sol* above the Amœba. This animalcule is a small microscopic Ileliozolarian abounding in fresh-water ooze. It casts out long, slender, filamentous pseudopodia from every part of its body. When its prey or any alimentary substance gets into the midst of this mass of filaments, the filament affected quickly draws back, carrying the nutritive matter with it towards the body proper of the Actinophrys. In other instances, the filaments, anastomosing themselves, form a sort of envelope about the prey. At the instant the substance comes within a short distance of the cellule, a part of the protoplasm composing the mass projects itself forwards, and encompasses the food, which is carried back and enveloped in the midst of the protoplasm by a process analogous to that seen in Amœba.

In the case of the Actinophrys any part of the body could serve as a way of entry for food, that is to say, could act the part of a mouth. To use the expression of W. Saville Kent, it is a *pantostomate* being. In other species of higher organization, this mode of alimentation is rendered impossible by the cutitcle which encompasses the body; the formation of a cuticle impervious to solid foods creates the necessity of a buccal orifice through which food may enter into the interior of the protoplasm.

A curious graduation in these phenomena is noticed here. Thus there are organisms destitute of a permanent and pre-existing mouth; their mouth is improvised as the occasion demands, is adventitions, so to say, and the reason that these organisms are ranked higher than the preceding ones, is that the mouth is invariably formed in the same place.

In this connection we may examine a small flagellate Infusory which abounds in impure waters, the *Monas vulgaris.* It carries a long flagellum attached to its anterior extremity, which when not in motion, is coiled up against the body. At the base of the flagellum the protoplasm projects a pellucid substance in the shape of a lip. This protuberance is hollow, containing a vacuole filled with liquid. Cienkotvski has described how these different organs act. The Bacteria and Micrococcus, which constitute the food of the *Monas*, are pulled into the latter's neighborhood by strokes of the flagellum; at that instant, the animal becomes conscious of the proximity of these other bodies, for the protuberance which lies at the base of the flagellum extends towards the corpuscule, envelops it in its own substance, and pulls it back into the interior of the Monad's body. Bütschli has made an analogous observation with the *Oikomonas termo.*

The prehension of food comprehends, here, three phases, in two of which the organism manifests psychical activity : *first*, attraction of food by means of the flagellum ; *second,* formation of the vesicle which extends towards and envelops the food, when the latter has come near; *third,* absorption of the food.

The Acinetæ are organisms that move about very little ; they frequently remain fixed to a pedicle their whole life long. They have no cilia, but exhibit radiating prolongations, more or less numerous, and sparse or grouped in tufts, as the case may be. These filaments are suckers, provided at the end with a small air-hole. When a thoughtless Infusory swims into the territory of an Acineta, the latter arrests it by means of its stout filaments and fastens upon the former's body the cup-shaped extremities of its suckers, which make a vacuum. The protoplasm of the Ciliate thus captured, slips slowly through the suckers as through tubes, and is gathered together in the interior of the Acineta in the form of small drops. In the Acinetæ, accordingly, particular organs are adapted to the prehension and absorption of food. Corresponding to the greater complexity of physical action, the psychical process necessary for the act of prehension has likewise become more complicated than is the case with the Amœba. The Acineta is obliged to *direct* its sucker towards the Infusory which is within its reach, and consequently the animal is obliged to determine the position of its prey.

There are Acinetidæ that exhibit prehensile organs more perfect than those just noticed. Such are the *Hemiophrys.* They have both sucker tentacles and prehensile tentacles. The latter are filaments which the animal throws about its victim like a lasso, thus enveloping and rendering it motionless, while it proceeds to feed upon it by means of its suctorial apparatus.

Now, do these Acinetidæ show any preference of choice among the Infusoria that chance to fall within reach of their tentacles? M. Maupas, who has made an especial study of these organisms had at first admitted this preference in choice. But he afterwards rejected the notion. In 1885, he writes us: "I find quite another explanation of the impunity with which the *Coleps hirtus* can throw itself upon the terrible suckers of the *Podophrys fixa*. The stout shell with which this little Infusory is enveloped, serves it as a shield and guards it from the deadly grasp of the Acinetidæ. The Acinetidæ do not seize the Coleps because of any dislike of the latter, but because they are unable to seize them, and their inability results from the peculiar structure of the Coleps' tegumentary envelope. The Paramecia which also escape unscathed, are similarly provided with a tegument of high resisting power, which serves them as a protection in this contingency. The *Stylonichia histrio*, like all other Stylonichiæ, has a very soft tegumentary envelope. They are accordingly seized and devoured by the Acinetidæ without difficulty. The detailed knowledge of the differences of structure in the tegumentary envelopes has caused me to abandon the idea of a preference or dislike in the choice of those victims which serve as food for the Acinetidæ. Of the prey that passes by, they catch what they can and not what they want to."

In a large number of species the prehension of food is preceeded by another stage, the search for food, and in the case of living prey, by its capture. We shall not investigate these phenomena among all the Protozoa, but shall direct our attention especially to the ciliated Infusoria. Their habits are a remarkable

study. If a drop of water containing Infusoria be
placed under the microscope, organisms are seen
swimming rapidly about and traversing the liquid
medium in which they are in every direction. Their
movements are not simple; the Infusory guides itself
while swimming about; it avoids obstacles; often it
undertakes to force them aside; its movements seem
to be designed to effect an end, which in most instances
is the search for food; it approaches certain particles
suspended in the liquid, it feels them with its cilia, it
goes away and returns, all the while describing a zig-
zag course similar to the paths of captive fish in
aquariums; this latter comparison naturally occurs to
to the mind. In short, the act of locomotion as seen
in detached Infusoria, exhibits all the marks of voluntary movement.

It might also be mentioned that every species
manifests its personality in its mode of locomotion.
Thus, as a rule, the *Actinotricha saltans* when placed
in a preparation where it finds itself at ease, remains
for a few moments perfectly immovable. Then, of a .
sudden, it dashes forward with the rapidity of light-
ning and disappears from the field of vision. For a
time it darts about to the right and to the left, and
then once more assumes its state of immobility. It
can move with the greatest agility through masses of
débris, in the midst of which, bending and twisting, it
slips about with wonderful nimbleness. The *Lagynus
crassicolis*, on the other hand, moves along at a pace
quite constant and uniform, neither slow nor rapid.
It searches about among algæ and fragmentary particles. The *Peritromus Emmæ* moves slowly. It runs
lazily over the Algæ, where it seeks its nutriment, and
does not stray from them to venture into the open water.

Concerning the prehension of foods and the search for nutriment on the part of Ciliates, we can do no better than to quote entire a note which M. E. Maupas has been pleased to send us upon the subject. We had put to him two questions: *First*, do the Ciliates hunt their food? *Second*, while in quest of live prey, do the Ciliates called hunters make an actual hunt, involving the espial of prey from a distance and the voluntary pursuit of the same in the circuitous paths they follow? M. E. Maupas after having once more had recourse to observation, briefly recapitulates his opinion in the following lines:

"From the standpoint of prehension of food, the Ciliates may be divided into two great groups:

1. Ciliates with alimentary vortices;
2. Hunter Ciliates.

"In the first group the mouth is always held wide open, and along with the nutritive particles which the current of the vortex keeps constantly sucking in, we may at will cause other, absolutely inert and indigestible, particles to take the same course; for instance, such substances as granules of carmine, indigo, and rice-starch. These granules, totally unfit for nutritive purposes, pass through the body of the Ciliates along with the genuine nutriment and are finally cast out intact with the excrement. I think, therefore, we may affirm that the species having alimentary vortices exercise no real choice in selecting their foods, and that they absorb indiscriminately all corpuscles which by reason of their form and density admit of being seized and drawn into the alimentary whirlpool.

"In the case of the hunter Ciliates proper, the mouth is constantly closed. The act of absorbing each object captured is accomplished by a process of de-

glutition comparable in every phase to the like process in higher animals. Furthermore, these species
feed only upon living prey, which they capture and
entrammel by means of their trichocysts (*vid. Archives
de Zoologie*, Vol. I. 1883, p. 607 and ff.). By this very act
they exercise a choice in the selection of food. But
this manifestation of choice is not, in my opinion, the
result of preference, or of individual taste, but is the
consequence of the peculiar construction of their buccal apparatus, which does not enable them to take
other and different nourishment.

"These hunter Infusoria are constantly running
about in quest of prey; but this constant pursuit is
not directed towards one object any more than another. They move rapidly hither and thither, changing their direction every moment, with the part of the
body bearing the battery of trichocysts held in advance. When chance has brought them in contact
with a victim, they let fly their darts and crush it; at
this point of the action they go through certain manœuvres that are prompted by a guiding will. It very
seldom happens that the shattered victim remains
motionless after direct collision with the mouth of its
assailant. The hunter, accordingly, slowly makes his
way about the scene of action, turning both right and
left in search of his lifeless prey. This search lasts a
minute at the most, after which, if not successful in
finding his victim, he starts off once more to the chase
and resumes his irregular and roving course. These
hunters have, in my opinion, no sensory organ whereby they are enabled to determine the presence of prey
at a distance; it is only by unceasing and untiring
peregrinations both day and night, that they succeed
in providing themselves with sustenance. When prey

abounds, the collisions are frequent, their quest profitable, and sustenance easy; when scarce, the encounters are correspondingly less frequent, the animal fasts and keeps his Lent. The *Lagynus crassicolis*, accordingly, never sees its victim from a distance and in no case directs its movements more towards one object of prey than towards another. It roams about at random, now to the right and now to the left, impelled merely by its predatory instinct—an instinct developed by its peculiar organic construction, which dooms it to this incessant vagrancy to satisfy the requirements of alimentation.

" The vorticel Infusoria, when in a medium abounding in food, are almost entirely sedentary in their habits, only making slight changes of position. But if they are placed in a medium affording but little nutritive material, they become as migratory as the hunters, and are seen to race about in all directions searching for more abundant nutriment. It is hard to find a more perfect illustration of the influence exerted by the conditions of a medium upon the habits and customs of animals.

"The *Leucophrys patula* is a type distinctively carnivorous and possessed of an extremely voracious appetite, a fact which explains its power of multiplication, one of the greatest I have studied. With a temperature of 25° in my laboratory I have recently seen it separate by fission seven times in twenty-four hours, that is to say, a single individual produces from itself just one hundred and twenty eight others in that time. In constant pursuit of its prey, it seizes its victims by the two stout vibratile lips with which its mouth is armed, and swallows them alive and whole. The victims may be seen struggling and tossing about

for a time in the interior of the Leucophrys's body
and afterwards to expire slowly under the action of
the digestive juices of the vacuole in which they have
been enclosed. Placed in a medium well-stocked with
small Ciliates, the Leucophrys have their bodies constantly crammed with victims swallowed in the manner above described. Like the other hunter Ciliates
the Leucophrys does not espy its victims from a distance and does not guide itself towards them. It
simply darts about from right to left, every moment
changing its direction. It thus increases its chances
of coming in collision with its prey and every time
that one of its unfortunate victims falls in contact
with its vibratile lips, it is seized, irresistibly drawn
towards the mouth and swallowed within less than a
tenth of a minute."   ·

Certain hunter Infusoria have methods of pursuit
and capture which deserve to be examined separately.
Claparède and Lachman in their excellent work upon
Infusoria and Rhizopods, have minutely described the
manner in which a large Infusory, the *Amphileptus
Meleagris*, attacks the *Epistylis plicatilis*. The *Epistylis* are colonizing vorticels of which certain individual members attain a size of not less than 0·21 mm.
The *Epistylis* form aborescent groups, the ramifications of which are quite regularly dichotomous. These
ramifications all grow at exactly the same rate and the
individual branches all rise to the same height, representing what is called, in botany, a corymbous in-
florescence. "We were observing one day," says
Claparède, "in the hope of seeing what would come
of the manœuvre, an *Amphileptus*, which was slowly
creeping upon a colony of *Epistylis*. The way in
which it approached the Vorticels, feeling them, so to

speak, and partly enclosing them in its pliable body, already seemed suspicious. At last, it made a direct attack upon one of them by fastening itself upon the upper part of its body. It opened its huge mouth, which is never to be seen except when the animal is eating, and slipped over the *Epistylis* like the finger of a glove being drawn upon a finger of the hand. We saw the sides of the buccal aperture (which are capable of being dilated in a truly astonishing manner) slip slowly over the peristome and upon the body of its prey, and then draw together about the point where it was made fast to the pedicle. The cilia covering the body of the *Amphileptus* began to shake with that peculiar motion which is always noticed when a ciliated Infusory secretes a cyst. At the expiration of a moment or so, a fine line was seen to appear around the whole body which continued to spread so as soon to form the cyst." (This might be called a cyst of digestion.) "The phenomenon as a whole is quite simple. An *Amphileptus* approaches an *Epistylis* devours it and encysts itself upon the spot, the victim being still attached to its pedicle. It then endeavors to wrench the *Epistylis* from its point of attachment by twisting; it turns on its axis from left to right and then from right to left, successively; when it has succeeded, it continues its work of digestion, and occasionally divides in two within the cyst itself. During the last stage of digestion, it rests for a while, when it commences again to turn about in the cyst, evidently seeking to disengage itself. At the close of a certain number of hours, the cyst breaks. The *Amphileptus* issues forth and starts in quest of another victim."*

---

* *Etudes sur les Infusoires et les Rhizopodes*, Vol. II. p. 166, 1861.

The hunter Infusoria are frequently armed with *trichocysts*. Trichocysts are urtical filaments which serve the animalcula provided with them to disable or wound other micro-organisms.

A large unmber of Infusoria, the *Paramecia*, the *Ophryogleuæ*, etc., use their trichocysts as organs of defense. With other species, of which we shall speak more at length, the trichocysts are organs of offense. They are located either in the sides of the mouth or in parts adjacent thereto; this is the case with the Lacrymaria, the *Didinium*, the *Enchelys*, the *Lagynus*, the *Loxophyllum*, and the *Amphileptus*.

These latter animalcula attack the live prey that constitutes their food, in the following manner. They dash upon their victim and bury the trichocysts with which they are armed, into its body. The victim is immediately brought to a halt, whereupon the hunter seizes it and swallows it. So, when the *Lagynus Elongatus* intends to seize a victim that has fallen into its vortex and has thus been drawn into the neighborhood of its mouth, it throws itself swiftly forward. At the moment of contact the hunted Infusory becomes suddenly paralyzed and remains perfectly motionless. This paralysis is evidently caused by the trichocysts which line the æsophagus of the Lagynus and with which the latter has transpierced its prey at the moment it came in contact by its anterior extremity.*

In a higher stage of organization, the Microzoön possessing a mouth changes its position in order to intercept its prey, and give it chase.

The *Didinium Nasutum* (Stein), a carnivorous Infusory and one of the most voracious of our fresh stagnant waters, operates in a more complicated manner:

---

* Maupas, op. cit., p. 495.

it casts its trichocysts upon its victim from a distance. The importance of this instance induces us to stop here a moment.

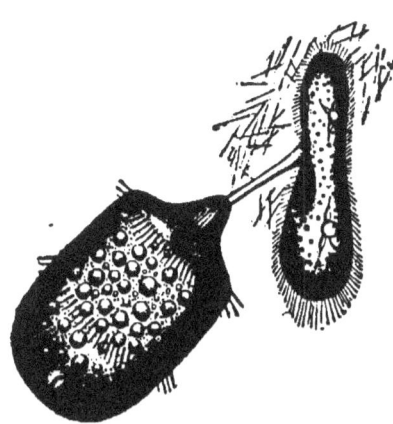

The *Didinium* (fig. 7), as regards the general shape of the body, may be compared to a diminutive cask, rounded off at one of the ends and terminated at the opposite extremity by an almost level surface from the midst of which rises a conical projection quite strongly marked. This projection is an organ of deglutition (swallowing); a longitudinal striation is noticed here formed of minute solid rods, of extreme tenuity and independent of the sides. These organs are the weapons used by the *Didinium* in attacking the live prey which constitutes its sole nourishment.

Fig. 7.—*Didinium nasutum*, enlarged two hundred diameters. The figure represents a *Didinium* overpowering a *Paramœcium aurelia*. The nettle-like filaments discharged by the *Didinium* are seen on all sides of the *Paramœcium*; while the latter, already seized by the tongue-shaped organ of the *Didinium*, is being gradually drawn towards the buccal orifice (after Balbiani).

Not only does it attack and devour animalcula almost as large as itself, but frequently it even seizes individuals of its own kind. In such cases it is always Infusoria, and never the Rotatoria, although the latter often abound in waters which the *Didinium* inhabits. It appears, moreover, to have a marked predilection for certain species; and so it happens that the huge and inoffensive *Paramœcium aurelia* is almost always its choice by preference among the animalcula that inhabit the same liquid.*

* The *Didinium*, Balbiani tells us, never attacks the *Parmœcium bursaria*, which is distinguishable from the *P. aurelia* by its green coloration.

The prehension of food by the *Didinium* exhibits interesting aspects, which have not as yet been observed in any other Infusory. M. Balbiani, in his first observations, had often been surprised at seeing animalcula that the Didinium had passed by without touching, suddenly stop as if violently paralyzed; whereupon our carnivorous specimen straightway approached and seized them with seeming facility. More careful examination of the Didinium's actions soon furnished the key to this enigma. If, while swiftly turning in the water, the *Didinium* happens into the neighborhood of an animalculum, say a Paramecium, which it is going to capture, it begins by casting at it a quantity of bacillary corpuscules which constitute its pharyngeal armature. The Parmecium immediately stops swimming, and shows no other sign of vitality than feebly to beat the water with its vibratile cilia; on every side of it the darts lie scattered that, were used to strike it. Its enemy then approaches and quickly thrusts forth from its mouth an organ shaped like a tongue, relatively long and resembling a transparent cylindrical rod; the free, extended extremity of this rod it fastens upon some part of the Paramecium's body. The latter is then gradually brought near by the recession of this tongue-shaped organ towards the buccal aperture of the Didinium, which opens wide, assuming the shape of a vast funnel in which the prey is swallowed up.*

Up to this point we have paid little attention to movements of defence and of flight. Upon this subject a few words will suffice. When vorticels are alarmed, they are seen to contract forcibly their pedi-

---

* *Archives de zoologie expérimentale*, 1873, Vol. II. p. 363. *Observations sur le Didinium nasutum*, by E. G. Balbiani.

cle, which in a state of rest stays extended. Infusoria placed in a preparation where they are at their ease, swim quietly about; if any sharp excitation disturb them, they accelerate their pace; those armed with a rigid bristle at the posterior extremity, rush precipitately onward whenever another Infusory chances to touch that tactile appendage. The unaggressive Parmecia, when attacked, endeavor to escape, but are also able to defend themselves by means of the trichocysts with which their ectosarc is armed.

IV.

Unicellular organisms do not all live in a detached state; a. large number of species are found grouped together in colonies; the initial basis of these agglomerations is always a mother cell, the offspring of which instead of dispersing to live at large, remain agglutinated to one another. Ehrenberg had believed that in certain species (especially in the case of the *Anthophysa vegetans*, an aggregation of minute monads growing as a sort of bush) the colony was created by the union of minute organisms that originally lived at large; but observation has shown that his theory was incorrect. It may be laid down as a general rule that every colony of monocellular animals or vegetables spring from the divisions of a single cellule. The cellules of one and the same colony, therefore, are always sister cellules, and the colony represents a family in miniature.

A leading instance of a colony wholly temporary, is found in those organisms the cuticle of which does not take part in the phenomena attending the division of the protoplasm. In this case, the protoplasm beneath the envelope .alone divides; the segments resulting therefrom are often numerous, and it is not until the plasma has finished dividing that the maternal cuticle is destroyed and that the segments separate to live abroad in a detached state. Up to that time they remain bound together.

It is thus seen that the existence of this minute colony is a transient phenomenon, which lasts only during the time necessary for the division of the maternal body. These phenomena have been noticed among many of the Flagellates. What appears surprising is, that the maternal cellule, although continuing to divide beneath the envelope, keeps on moving about in the water by means of its own flagellum as if still constituting only a single animal. The reason of this is that one of the segments into which the plasm is divided and which is situated in the anterior part of the mother-cellule, remains connected with the flagellum and takes charge of its movements. This segment (like an individual distinct in itself) alone guides the bark that carries its sisters. And so, although this diminutive colony is as a rule but short-lived, a division of labor has been effected among its members; the anterior segment is alone entrusted with the office of locomotion.

The colony has a duration less ephemeral in the case of the *Gonium pectorale*, a Volvocine known in our fresh waters. It is formed by the aggregation of sixteen individuals which remain detached but adhere laterally to one another. The colony is developed in one way only: it is in the form of a minute rectangular plate of a beautiful green color. In the case of the *Pandorina*, the colony assumes the form of a minute sphere; it is composed of sixteen, or as many as thirty-two individuals, joined together beneath a stout envelope; each member remains free in action, and projects its two flagella through the cuticle. With the *Eudoryna elegans*, the colony is modeled upon nearly the same plan excepting that it is composed of thirty-two individuals and that the latter, placed beneath the same cuticle at equal distances apart, do not touch one another.

In the genus Volvox, colonies are found of which the structure is very complicated. Such are the great green balls formed by the aggregation of diminutive organisms, which form the surface of the sphere, and are joined together by their envelopes; they have each two flagella, which pass through the enclosing membrane and swing unimpeded on the outside; the envelopes, each tightly holding the other, form hexagonal figures exactly like the cells of a honeycomb. Each Volvox is at liberty within its own envelope; but it projects protoplasmic extensions which pass through its cuticle and place it in communication with its neighbor. It is probable that these protoplasmic filaments act like so many telegraphic threads to establish a network of communication among all the individuals of the same colony; it is necessary, in fact, that these diminutive organisms be in communication with each other in order that their flagella may move in unison and that the entire colony may act as a unit and in obedience to a single impulse. The number of micro-organisms constituting a Volvox colony is quite considerable: as many as 12,000 have been counted.

It was upon analogous phenomena that Gruber based the existence of a diffused nervous system in the Stentors. The same line of reasoning may be followed in the case of the Volvox. Since unanimity of movement is demonstrable among twelve, thousand micro-organisms constituting a colony, it must be inferred that their movements are regulated by the action of a diffused nervous system present in the protoplasm. This conclusion is all the more interesting from the fact that these Volvox are vegetable micro-organisms.

In the diœcian Volvox, the female cellules and the male cellules are joined together by themselves in separate colonies. When the time of fecundation arrives, the male cellules or antherozoids scatter and proceed to conjugate with the female cellules. The colony which bears the female cellules also contains neutral cellules which are not designed for fecundation; the latter simply perform a locomotive function; equipped with one eye and two flagella, they are intended to move the great colonial ball: they are the oarsmen of the colony. The Volvox, male, female, and neutral, all seek the light, whether solar or artificial, and settle near the surface of the water. As soon as the female colonies have been fecundated, the oöspores change their color: they turn from green to an orange yellow. At this point, the colony is seen to draw away from the light and to disappear from the surface of the water. This change of position is effected by means of the vibratile cilia with which each neutral cell is furnished and which project beyond the gelatinous sphere; now, as no change of color or form is noticed in the neutral cells after fecundation, it may be asked from what cause they flee from the light which they formerly sought.

Colonies of Proto-organisms formed by the division of a mother cell of which the segments remain united, are not entirely without analogy with a pluricellular organism which likewise springs from a single cell called the egg, and the resultant divisions of which do not separate.

The colony constitutes in a way a first step towards the physiological constitution of a pluricellular organism; it serves to fix a stage of transition in the animal kingdom, between Protozoa and Metazoa. A fact which strengthens this analogy is, that certain colonies, as the *Synura uvella* and the *Uroglena volvox*, can divide into two other colonies; strangulation acts upon the mass just as if upon a pluricellular organism. This curious observation was made by Stein and Bütschli.

Nevertheless, an essential difference still separates the Metazoa and the Protozoan colonies, even when in these colonies a division of function has been established among several individual groups. The physiological differentiation brought about in these Protozoan colonies is the result of a mechanism which differs in every respect from that by which it is effected in the case of the Metazoans. In the latter instance the differentiation results from the division of the embryo into *germinative folia* each of which is the origin of a separate group of organs. At a certain stage of development, the superposition of these folia gives rise to the formation of a *gastrula;* the *gastrula* is formed by two folia joined together, representing a pouch open to the outside; it is characteristic of Metazoans, the Protozoan never reaching this stage. Certain colonies observed by Hæckel, the *Magosphæra planula* for example, and the volvox, of which we have before spoken, appear in the form of a sphere; they suggest an anterior stage of development to which the name of *morula* or of *blastula* has been given; but they do not get beyond this stage.

We have now considered assemblages of organisms which live joined together like the Gonium and sometimes united by a material band like the Volvox, where the individuals are grouped together under one and the same cuticle. Voluntary and free combinations are much more rarely met with; nevertheless cases occur. There exist organisms which lead a life of habitual isolation but which understand how to unite for the purpose of attacking prey at the desired time, thus profiting by the superiority which numbers give.

The *Bodo caudatus* is a voracious Flagellate possessed of extraordinary audacity; it combines in troops to attack animalcula one hundred times as large as itself, as the Colpods for instance, which are veritable giants when placed alongside of the *Bodo*. Like a horse attacked by a pack of wolves, the Colpod is soon rendered powerless; twenty, thirty, forty *Bodos* throw themselves upon him, eviscerate and devour him completely (Stein).

All these facts are of primary importance and interest, but it is plain that their interpretation presents difficulties. It may be asked whether the Bodos combine designedly in groups of ten or twenty, understanding that they are more powerful when united than when divided. But it is more probable that voluntary combinations for purposes of attack do not take place among these organisms; that would be to grant them a high mental capacity. We may more readily admit that the meeting of a number of Bodos happens by chance; when one of them begins an attack upon a Colpod, the other animalcula lurking in the vicinity dash into the combat to profit by a favorable opportunity.

v.

It is difficult in the extreme to mark out the lines of a psychology of Proto-organisms from data so incomplete as those we have just collected. We shall content ourselves with a few brief considerations.

The apparent result of our investigations up to this point is, that the greater number of movements and actions observed in Micro-organisms are *direct* responses to excitations emanating from the medium in which they live. It is the condition of the medium that, to all appearance, rigidly determines the character and manner of their activity; in a word, they exhibit no marks of pre-adaptation.

But it will not do to let the matter rest with this general survey of the subject; we shall have to examine more closely each detail of these reflex actions of adaptation, beginning with the sensory phase and ending with the motory phase. Analysis discloses that several determining elements may be distinguished in these phenomena; they are:

1. The perception of the external object;
2. The choice made between a number of objects;
3. The perception of their position in space;
4. Movements calculated, either to approach the body and seize it, or to flee from it.

We are not in a position to determine whether these various acts are accompanied by consciousness or whether they follow as simple physiological processes. This question we are obliged, for the present, to forego.

1. *The perception of an external body.* Among the lowest forms, it appears that perception is always the result of a direct irritation produced by contact of the external body with the protoplasm of the animalcule. This is what takes place, to all appearance, among the

Amœbæ; for these organisms, the condition necessary to the perception of a solid particle is contact with it. A step forward has been effected in those organisms that are able to perceive external objects by contact from a distance, as is observed for instance in the *Actinophrys*, which perceives all bodies that chance to touch its long filamentous pseudopods; yet, in this instance, the pseudopod merely acts the part of an extended tactile organ. The vibratile cilia, and still more the long lash of the Mastigophores, enable the animal to discern the presence of contiguous particles at a certain distance from its body, by the pressure exerted upon their appendages. It is not known whether there are many animalcula that perceive the presence of nutriment from a distance and without coming in direct contact with it; it appears, however, that this is the case with the *Didinium* which shatters its prey from a distance and without touching it.

2. *Choice.* We have seen that Micro-organisms do not absorb indiscriminately every solid particle they meet. They exercise a choice. Among the lower species, the choice is in the lowest degree rudimentary; the organism restricts itself to a discrimination of mineral particles, sand for example, from organic substances; it rejects the former and absorbs the latter. Among the higher animalcula the choice is more intelligent. There are Infusoria that feed only upon plants and animals. There are also those which feed exclusively upon one species.

This exercise of choice is one of the most incomprehensible of phenomena; it is exceedingly difficult to explain it without resort to anthropomorphism. If we hold to what observation directly teaches us, the choice may be said to consist in the following acts:

when the animalcule perceives certain kinds of substances and particularly those substances which serve
it as customary food, it invariably goes through the
same movement, which consists of an act of prehension;
when the substance touched, seen, or collided with,
as the case may be, is of another kind, the Micro-organism does not go through this act. Such is the
phenomenon; as to the explanation of the same, we
are unable to give one.

According to M. E. Maupas, if certain Infusoria
feed exclusively upon a certain species, it is because
their buccal apparatus, or organ of prehension, makes
it impossible for them to feed upon different species
which possess different tegumentary envelopes. The
question is to ascertain whether this explanation is
applicable only in certain cases, as appears very probable to us, or whether, on the other hand, it is of complete and universal applicability. We confess that
the hypothesis of M. Maupas does not explain to us
why a hunter Infusory that throws trichocysts, like
the Didinium, attacks the *Paramœcium aurelia* and not
the *Paramœcium bursaria.*

It is possible that certain species attract the organisms which feed upon them, by means of a physical or chemical excitation.

The researches of Prof. Pfeffer, of the Tübingen
Botanical Institute, lend a certain confirmation to this
hypothesis.

3. *Calculation of the position occupied by the external body.* It is a universal fact that Micro-organisms
not only perceive external bodies, but that they also
indicate, by their movements, an exact knowledge of
the position occupied by these bodies. It might be
said that they invariably possess a sense of position in .

space. The possession of this sense is absolutely indispensable to them, for it does not suffice them to know of the presence of an exterior body in order to approach it and seize it; they must furthermore know its position, so as to direct their movements accordingly.

The simplest form of a sense of localization is met with in the Amœba, which, when it closes about a nutritive particle, always emits its pseudopods at precisely that part of its body where the foreign substance caused the irritation. The most complicated instance of localization is met with in the *Didinium*, which we have so often cited; the *Didinium* knows precisely the position of the prey it follows, for it takes aim at the object of its pursuit like a marksman, and transpierces it with its nettle-like darts. Between these two species, we find all the intermediate instances of a localization of perceptions.

However, doubts exist upon the question as to whether Proto-organisms know the direction and distance of external bodies, or whether they only succeed in getting at them after a series of tentative movements. The observations which we have collated do not solve the question.

4. *Motory phase.*—We now pass to the motory phase. The movements made by Micro-organisms as if in response to an excitation, are not in most instances simple reflex motions; they are movements adapted to an end. We cannot repeat it too much: these movements are not explained by the simple phenomenon of cellular irritability.

In the very first instance, they vary according to the excitation; a given excitation produces a corresponding motory response; a body situated at the right does not bring about the same movement that a body situated at the left does; a particle of the nutritive sort does not provoke the same course of action that a particle of a different sort does. All this implies that associations have been established in the protoplasm between certain excitations and certain movements. The explanation of the physical nature of these association appears to us totally impossible.

The quite ingenious ideas broached by Spencer upon the lines of least resistance offered by the commisural fibres cannot be applied here, since everything takes place in a single cell. What would be necessary to explain is how and in consequence of what mechanism of structure one form of molecular movement, corresponding to a given excitation, is followed by a certain other form of molecular movement corresponding to an act likewise determined.

VI.

FECUNDATION.

We now enter upon a subject fraught with obscurity. We shall limit our investigations to ciliated Infusoria, as it is among these species that fecundation and the psychical phenomena attendant thereon have been best observed.

Ehrenberg had established by his authority the prevailing opinion in science that copulation never takes place among Infusoria, and that all facts observed by early writers as connected therewith are to be regarded as phenomena of longitudinal fissiparity. This erroneous idea prevailed unquestioned until 1858, when M. Balbiani addressed a communication to the Academy of Sciences, wherein he showed that sexual reproduction, preceded by copulation, *is* found among Infusoria.

Before entering upon a description of the changes that take place in the nucleus and nucleole of Infusoria in coition, we shall briefly sketch the course of psychical phenomena through which the ciliated Infusoria pass when making ready for copulation.

We shall follow in the footsteps of M. Balbiani, freely using his descriptions, the exactitude of which has since been confirmed by Gruber.

To appreciate fully the significance of the facts to be adduced herewith, we must recall to mind that throughout the entire animal kingdom the act of sexual coition is invariably preceded by an introductory manifestation of psychical activity, which may last for quite an extended length of time.

The female, when pursued by the male, seems to be animated by two conflicting desires—that of yielding to the male and that of repelling his approaches. This show of unwillingness, which is but temporary and more seeming than real, has the effect of inciting the male to attempt an exhibition of powers calculated to captivate the female. According to M. Espinas, who has thoroughly studied this subject, there are five classes of phenomena which assist in preparing the way for sexual union: firstly, provocative contact, the lowest of all these phenomena—that is, the one which most approximates to the physiological order; secondly, odor; thirdly, color and form; fourthly, noise and sound; fifthly, play, or every variety of movement. It appears to us that almost all manifestations of love in human beings themselves could be classified into these five categories.

Among the simplest forms of life we meet with in-

cipient traces of such æsthetical manifestations point-
ing towards the preparation of two animals for sexual
intercourse.

" It is curious," remarks M. Balbiani, " to find
among these organisms which all zoölogists, by reason
of their diminutive size and extreme simplicity of
structure, have placed at the remotest limit of the animal
kingdom, acts that mark the existence of phenomena
analogous to those by which the sexual instinct is exhibited in a large number of Metazoans. Upon the
approach of the period for propagation, the Paramecia
come in from all points of the fluid and assemble like
little whitish clouds in more or less numerous groups
about the objects that float upon the surface of the
water, or adhere to the side of the vessel containing
the tiny artificial sea in which the animalculæ are held
captive. Intense excitement, which the need of food
does not suffice to explain, prevails in each of these
groups; a higher instinct appears to dominate all these
tiny organisms; they seek each other's company, chase
each other about, feel here and there with their cilia,
adhere for a moment or so in an attitude of sexual coition, and then retire, soon to begin anew. When
these minute assemblages are dispersed by shaking
the liquid, they quickly form again at other points.
These singular antics wherewith animalculæ appear
to incite each other mutually to copulation often
continue for several days before the latter act is definitely effected.

" Other Infusoria, particularly the Spirostomes, seek
the deep spots of the liquid, or bury themselves in the
oozy sediment of the bottom, not to come forth again
until they have separated. The Stentors have different habits. They are affixed by their pedicles to sub-

merged vegetable patches, which they often cover like small, closely-mown lawns, of a green, brown, or blue color, according to the species; they turn the forward part of their bodies, which is elongated in the shape of a trumpet, about in all·directions, and seek to unite with each other by the broadened extremity which corresponds to the bell of the trumpet."

Among the numerous species forming part of the group of Oxytrichinæ, the act of coition likewise exhibits certain interesting preliminaries. The two individuals, whose bodies are generally very much flattened, and of which the lower sides are provided with cilia at times strongly developed, superpose themselves upon each other on the ventral side and mutually entangle the cilia which cover that region, while with their cornicles, or anterior tentacles, they touch repeatedly the different parts of each other's bodies. These introductory moves frequently last for several hours before copulation begins.

As regards the act of copulation itself, it too is of exceeding interest to the psychologist, who can admire the precision with which the two individuals assume the attitude necessary for fecundation.

During conjugation the two ciliated Infusoria are always joined together at the aperture which forms the mouth. It has been thought that this aperture must play the part of a sexual orifice through which the two animalculaº in copulation effected the exchange of reproductive matter; it has been suggested, moreover, that an especial sexual orifice was present, quite close to the mouth; but these questions of structure are still doubtful.

The attitude of these organisms during copulation

varies according to the position of the mouth which in certain groups is lateral and in others terminal.

The greater number of species have a lateral mouth. To this class belong the Paramecia; these Infusoria, in which the buccal fosse lies at the bottom of a deep excavation made in the ventral face, cover each other over the whole extent of this face, exuding a glutinous substance which causes them to adhere in this position; the two mouths then lie exactly upon each other. Copulation lasts from twenty-four to thirty-six hours with the *Paramæcium aurelia;* it lasts several days (five or six) with the *Paramæcium bursaria.* Among the Oxytrichinæ, the two animals in conjugation blend together at an important part of their persons in a very intimate fashion.

We next arrive to the second group of Infusoria, which show a terminal mouth; of this type we have had a specimen in the *Didinium nasutum*, the curious hunter Infusory; we may further mention the Coleps, the Nassula, the Prorodon. The two organisms, in this case, do not embrace laterally, they take a position end to end, connected by their anterior extremities, mouth opposite to mouth; then, little by little, while still joined at the buccal extremity, they shift about until they meet length to length.

We shall mention particularly, but briefly, the curious phenomena that accompany fecundation among the Vorticels. Even more than in the instances just cited do these phenomena resemble the process of fecundation in higher animals, for in this instance fecundation is effected between two differentiated individuals, one of which acts as a male element and the other as a female element. The Vorticels are colonies of Infusoria in which are found sedentary individuals, having the shape of minute jugs, and also detached individuals called Microgonidia, which are formed by repeated divisions upon the colonial tree.

These Microgonidia have exactly the same mode of locomotion as the spermatozoids. Engelmann * has followed their movements. He has seen them swimming about turning upon their axis for five or six minutes; then, having come into the vicinity of a Vorticel, they abruptly change their manner of movement, capering about the latter like a butterfly flitting about a flower, touching it, retreating, and then approaching it again and apparently feeling of it; at last, after having visited the others near by, they return to the first one and fasten themselves upon its surface. The coition is not effected without a certain show of resistance on the part of the Vorticel. It hastily contracts the peduncle to which it is attached, at every touch of the Microgonidium, while the latter in order to prevent itself from being thrown back by these rapid shocks and in order to be always close to the individual with which it wishes to unite, fastens itself by an extremely fine filament to the style of the Vorticel; thus attached and drawn along with the movements of the latter, it finally succeeds in effecting a junction with it and in penetrating into its body.†

It is now time to describe the material phenomena that take place in the interior of the two Infusoria, and which constitute the material act of fecundation. The psychical manifestations which we have just noted and which so strikingly resemble the manifestations accompanying the copulative act in higher animals, are of themselves sufficient evidence that this conjugation is a sexual union.

* *Arch. de Zoölog. expérimentale*, Vol. V, 1876.
† *Journal de Micrographie*, 1882, p. 241.

The material changes effected inside the bodies of Infusoria in copulation do not extend to all their organs; the main mass of the body, the protoplasm, plays but a secondary *rôle* in the matter; the change appears to be effected exclusively in the nucleus and the nucleole.

Let us further state that, so far as is known, these changes are never effected apart from coition and before the Infusoria actually embrace; copulation sets in every time, apparently, that these animals, under particularly favorable conditions, have actively reproduced by fission. Fissiparity is then seen to cease and conjugation appears.

We have not the time to sketch the history of this important question of physiology, interesting as it may be. It will be enough to recapitulate what we actually know upon the subject, taking as our guide substantially the views of M. Balbiani who, as is known, was the first scientist to study the physical phenomena connected with fecundation among Infusoria. The divergencies between his observations and those of another eminent investigator, M. Bütschli, extend in reality only to points of detail.

Let us first mark the modifications that take place within the *Chilodon cucullus* during conjugation. Each of the two Infusoria in copulation possesses a nucleus (endoplast, main nucleus) and, close beside this nucleus, an organ considerably smaller, a nucleole, or attendant nucleus, or latent nucleus (endoplastule, accessory nucleus); this minute body must not be mistaken for the nucleole that is often found in the interior of the nucleus among many Micro-organisms and in cellules; it has a function entirely different.

Of these two elements, the nucleus plays an almost negative part in the act of fecundation. It assumes irregular outlines and becomes rumpled, while its contents collect in detached masses of various sizes: it grows clear by degrees and is finally absorbed. It disappears, accordingly, by a phenomenon of regression and without dividing.

Fecundation aims to replace this wasted element by a nucleus of fresh formation. The latter is produced at the cost of the little body we have described by the name of attendant nucleus or latent nucleus. The attendant nucleus does not act in making up a main nucleus in the cellule of which it is a part; it finds its way into the body of the other animal and it is in this new cellule that it is destined to perform the function of a nucleus.

In the *Chilodon cucullulus*, the attendant nucleus divides into two striated capsules, never more. These two capsules grow to unequal sizes; the largest attains a size of forty thousandths of a millimetre; it is this one that forms the new nucleus of the Chilodon. The second capsule shrinks and becomes compressed; it takes its place beside the first one and constitutes the new attendant nucleus.

To the study of this type of fecundation we may limit our attention; it is the simplest of all, and other forms may be comprehended within it without much difficulty. What complicates the process in the other species is principally the successive modifications through which the old nucleus passes before suffering absorption. In the *Stentor cœruleus* the nucleus has the shape of a long chaplet or string of beads; at the moment of fecundation the beads of the chaplet break apart and spread in the protoplasm where they finally become absorbed. Among the Paramæcia the phenomenon is still different: the nucleus, at first massed together in a cluster, lengthens out into a very long string, which breaks; and the pieces becoming scattered about in the protoplasm, are absorbed.

We find that fecundation in every instance introduces the dispersion and disappearance of the old nucleus and that the latter is replaced by a new nucleus resulting from the transformation of the attendant nucleus that proceeded from the other organism.

The various modifications presented by this attendant nucleus likewise contribute in great measure to the complexity of the phenomenon. We have seen that in the Chilodon the attendant nucleus breaks into two globules, of which one goes to form the new nucleus and the other the new attendant nucleus. Matters take a different course in the Paramæcia. In the *Paramæcium bursaria*, for instance, the attendant nucleus divides into two and then into four capsules; one of these capsules suffers absorption, a second one becomes the attendant nucleus, and the two others coalesce with what remains of the old nucleus to form the nucleus proper. In the *Paramæcium aurelia* the division is made into eight capsules; three are cast out, and of the five left four are meant to form the new main nucleus; in reality, each Paramæcium segmentates first into two and then into four divisions, and each of these four individuals takes one of the capsules. The fifth capsule is designed to form the attendant nucleuses of these four organisms; it divides, accordingly, into two and then into four parts; that is to say, into as many parts as the body of the animal divided.

There is no question in our mind but that conjugation in this case is a sexual phenomenon. A circumstance that at the outset confirms this is the peculiar manœuvering the animalcula·go through before abandoning themselves to copulation; the movements they execute admit of exact comparison with the actions attendant upon copulation among higher animals. But we shall recur further on to the physiological significance of conjugation, when we shall endeavor to explain, according to the most recent investigations, the function of the nucleus in the cellule.

The question may be asked, what is the starting-point, the provocative of these sexual phenomena, the cause that sets them in play. Bütschli justly thinks, that conjugation is determined by internal causes; in fact, it takes place directly after very active periods of spontaneous division, as Balbiani has shown. When we bear in mind that the object of conjugation is to replace the old nucleus which has become wasted and worn out, we may conjecture with some degree of likelihood that the physiological condition of the nucleus constitutes the sexual excitant that causes the Infusoria to copulate.

However that may be, a curious observation with the *Paramæcium aurelia* has made us acquainted with one of the structural conditions of the sexual instinct in that Infusory. For a long time J. Müller had pointed to the presence of filaments in the nucleus and even nucleolus of Paramæcia, that had the appearance of spermatozoids. Observations to the same effect have increased since then, and it is now known that the filaments are Schizomycetes, parasitic Bacilli, which find their way into the nucleus and nucleolus, and multiply, after their customary mode of segmentation, by disarticulation. Balbiani has definitely determined the nature of these filaments by morphological and micro-chemical methods; he has found out, among other things, that the filaments do not dissolve in strongly concentrated alkaline solutions; and it is known that Bacteria exhibit this peculiar attribute of offering a great resistance to destructive agents.

In the nucleus which they have penetrated, these parasites induce a pathological condition that results in destroying every manifestation of the sexual instinct in this Infusory. Among a swarm of animals of this species that are in copulation, single individuals are found that show a nucleus and nucleolus comcompletely charged with Bacteria; sometimes these organs suffer an enormous dilatation, the nucleus becoming nothing more than an enveloping membrane which is filled like a huge pouch with parasites. The animal continues to live, but it no longer attempts to copulate.

VII.

It is not our intention to make a full and complete study of fecundation in higher animals and plants; there is but one phase of that phenomenon that can enter into a general study of Micro-organisms, and that is the history of the sexual elements, of their form, their movements, and lastly their copulation.

We shall describe animal fecundation first, and plant fecundation afterwards; regarding these phenomena particularly from a psychological standpoint.

Among metazoans, fecundation may be divided into two distinct acts. The first, and most apparent, consists of the union of the two individuals; of this we shall not have to speak here; it is a phenomenon that lies outside the limits of our investigations. The second, more deep-seated, consists in the phenomena that take place, after copulation, between the spermatozoid and the ovule.

There are numerous reasons for comprehending a study of the generative elements within a general investigation into the nature of Micro-organisms. In the very first place, it must be taken into account, that these two elements are represented in animals by a single cell.

The ovule appears as a minute microscopic sphere enclosed by an envelope (vitelline membrane); it is formed of a mass of granulous protoplasm (vitellus) containing a nucleus (germinative vesicle) and one or many nucleoli (germinative spot). The spermatozoids, in vertebrates, have quite a different aspect: they are filaments of varying lengths, having a distended part, or head, and a tapering, attenuated part, or tail.

The resemblance that spermatozoids bear to Protista, at first caused them to be regarded as animals living a parasitic life in the spermatic fluid. Ehren-. berg classed them among the polygastric Infusoria. Kœlliker and Lallemand were the first to reject this notion and the first to regard spermatozoids as elemental parts of living tissues, having the morphological value of a cellule. They are now likened to detached cellular elements, such as blood-globules.

Whatever form they assume, the sexual elements live as minute organisms independent of the individual from which they originated. This circumstance is particularly remarkable in the case of the male element, the spermatozoid, which retains its vitality for a certain space of time after its expulsion. The length of this period varies with the different species. Whereas the spermatozoids thrown from a trout lose all motion in the water after the expiration of a few seconds, those of the bee, in the seminal reservoir of the female, remain alive for several years. The seminal elements of mammifers live for quite some time in the genital passages of the female. Balbiani has found living spermatozoids in the ducts of a she-rabbit twenty hours after coition. Ed. van Beneden, Benecke, Eimer, Fries, have observed that the sperm retains its properties in the uterus of bats for several months.

Another remarkable circumstance is, that the copulation of the two sexual elements is not without analogy to the copulation of the two animals from which they originated. The spermatozoid and the ovule, to some extent, repeat on a small scale what the two individuals perform in their larger sphere. Thus, it is the spermatozoid that, in its capacity of male element, goes in quest of the female. It possesses, in view of the journeys it has to make, organs of locomotion that are lacking in the female and are useless to it. The spermatozoid of man and of a great number of mammifers is equipped with a long tail, the end of which describes a circular conical movement, which together with its rotation about its axis, determines the forward motion of the spermatozoid. The same mode of progression is seen in the zoöspores of Algæ and in Mastigophores, which are armed with flagella; the movements of the spermatozoid have been not improperly compared to those of a Flagellate.

Other spermatozoids like those of the Triton and Axolotl are provided with a different kind of locomotive apparatus; it consists of an undulatory membrane that acts like a real fin; the spermatozoid moves forward without turning about on its axis.

There has been much discussion as to the nature of the forces that account for the movements of the fecundative elements. The early investigators that concerned themselves with the study of animalcula, naturally attributed to them spontaneous and voluntary movement. Since the spermatozoid has been regarded as nothing else than an histological element, endosmotic, hygroscopic and like actions have been accepted in explanation. M. Balbiani, from whom we have taken the foregoing details, declares that explanations of this character are none at all; for, upon ultimate analysis, all kinds of motion may be reduced to a chemical or physical action—sarcodic or ciliary movement just as much as voluntary movement. "For my part," our scientist adds, "I believe that the spermatozoids do not move about blindly but that they act in obedience to a kind of internal impulsion, to a sort of volition which directs them towards a definite object."* The experiments of M. Balbiani have shown that with weak solutions of ether and chloroform the movements of the spermatozoids may be moderated and made to cease so slowly that the latter are yet able to fecundate the ovules.

In fine, the spermatic element, in directing itself toward the ovule to be fecundated, is animated by the same sexual instinct that directs the parent organism towards its female.

In the higher animals, the movements of the spermatozoid that is endeavoring to reach the female exhibit a peculiar character, which it is important to emphasize: these movements do not appear to be directly provoked by an exterior object, as those of micro-organisms are; the spermatozoid endeavors to reach an ovule which is frequently situated a great distance away; this is the case particularly

---

* La Génération des Vertébrés, p. 159.

with animals that fecundate internally, with birds and mammifers. The place of fecundation is still imperfectly known. Coste at one time accepted the theory that the spermatozoid and ovule met in the ovary. Fecundation probably takes place in the fore part of the oviduct. It has little to do with our purpose, however, to solve this delicate question precisely. A fact that is important to mention in a general way is the length of road the spermatozoid has to traverse before coming up with the ovule.

Let us now follow the spermatozoid in its journey to the ovule. It is known that the road it has to traverse is, in certain instances, extremely long. Thus, in the hen the oviduct measures 60 centimeters, and in large mammifers the passages have a length of from 25 to 30 centimeters. We might ask ourselves how such frail and minute creatures come by a power of locomotion great enough to enable them to traverse so long a path. But observation discloses the fact that they are able to overcome obstacles quite out of proportion to their size. Henle has seen spermatozoids carry along with them masses of crystals ten times larger than themselves, without appreciably lessening their speed. F. A. Pouchet has seen them carry bunches of from eight to ten blood-globules. M. Balbiani has attested the same fact. These globules, which have fastened themselves about the head of the spermatozoid, have each a volume double that of the head. Now, according to Welcker, the weight of a globule of human blood is 0.00008 of a milligramme: allowing that the spermatozoid has the same weight, we may then say that it is able to carry burdens four or five times heavier than itself.

The length of road traversed is not the only remark-

able circumstance here; there are also involutions and
intricacies in the path to be followed in reaching the
ovule. In this connection an interesting observation
has been made upon the silk-worm. "At the moment
of conjugation the male deposits its seminal fluid in a
special sac, the copulatory sac. The day following,
this sac, which was distended by the sperm, is completely flaccid, and nearly all the spermatozoids have
traveled out into another sac, which opens into the
oviduct opposite the first one, and there they wait to
fecundate the ovules as they pass by. Now, the walls
of the copulatory sac have no contractile power, and
the passage of the spermatozoids from one sac into
the other can be attributed only to a spontaneous
movement. Further, a fact that well seems to verify
this, is, that there still remains in the copulatory sac
a few misformed seminal elements, deprived of the
power of locomotion." *

Let us now note what happens at the moment
when spermatozoid and ovule come in contact with
one another. The successive phenomena then taking
place have been carefully studied by Fol in his work
upon the star-fish (*Asterias glacialis*). The ovule has
no enveloping membrane; it it is covered about only
by a mucous layer, soft and flaky. The spermatozoids
come up in great numbers and push forward into this
layer; at this point they are all brought to a halt and
become entangled among each other with the exception of one, which, more speedy in its movements, out-
strips the others and arrives within a short distance of
the surface of the vitellus (or protoplasm of the ovule).
At that moment, and before any contact whatever, there
results a curious phenomenon of attraction between the

---

* Balbiani, *Comptes Rendus de l'Acad. des Sciences*, 1869.

ovule and the spermatozoid; the peripheral substance of the ovule is seen to lift itself up in front of the spermatazoid in the shape of a minute protuberance; this protuberance, at first, has a rounded shape, then it grows thinner and forms a point which advances towards the spermatozoid; this point is called the cone of attraction (see Fig. 8). The head of the spermatozoid fastens itself upon the cone, which seems to draw it into its interior. The tail of the spermatozoid does not appear to enter into the interior of the ovule and take part in the process of fecundation, which consists simply in the fusion of the head of the spermatozoid with the nucleus of the cellule.

Fig. 8.—A small portion of the ovule of a star-fish (*Asterias glacialis*) showing the formation of the cone of attraction. (According to Fol.)

As soon as the head of the spermatozoid has penetrated into the ovule, the latter enwraps itself in an envelope, to protect itself against the other male elements. It appears, in fact, to be well settled that the penetration into the vitellus of several spermatozoids marks the beginning of an adverse change: the subsequent segmentation of the ovule is irregular, and development ceases.

The membrane in which the fecundated ovule of the *Asterias glacialis* infolds itself, is formed by a condensation of the peripheral layer of the vitellus; the condensation starts from about the point where the spermatozoid penetrated, and gradually spreads over the whole surface of the ovule; the formation of this protective membrane is so rapidly effected, that access to the ovule is barred against spermatozoids who might be only a few seconds behind the first one.

Sexual selection, then, acts among spermatozoids

just as among all animals; it is the most agile and the stoutest spermatozoid that first penetrates the ovule and effects fecundation. The laws of selection, thoroughly developed by Darwin, do not only apply to individuals; they apply also to sexual elements.

We are unable to follow the successive modifications suffered by the head of the spermatozoid after its entrance into the ovule; we may state simply, that the head presents the appearance of a radiate figure, of a diminutive sun advancing towards the female nucleus. At the same moment, the female nucleus appears affected and puts itself in motion towards the spermatic nucleus. The two nuclei soon come almost within contact, and it is in particular the female nucleus that then plays the active part. It is disturbed by incessant movements and every moment changes its form; it thrusts out prolongations towards the male nucleus, and one of these prolongations fastens itself upon the latter, presenting at the end a minute depression in the shape of a cup, which receives the male nucleus; and the two nuclei, while executing active movements, fuse into one another. In this manner the first nucleus of segmentation is created.

Selenka has furnished interesting chronological data as to the time of appearance of the different phenomena. The time is in each case taken from the moment of artificial fecundation. After a lapse of five minutes, the spermatozoid has forced an entrance into the ovule.* At the expiration of ten minutes (that is, five minutes after entrance), it has reached the centre of the ovule. At twelve minutes, the female nucleus has put itself in motion to meet the spermatic nucle-

---

* M. Balbiani, *Cours sur la fécondation*, passim. *Journal de Micrographie* Vol. III. 1879.

us. Finally, at the twentieth minute, the two nuclei have united.

In the psychical history of animal fecundation as just given, there are many gaps: the history of vegetable fecundation will fill several of these. ·

The simplest forms of sexual reproduction in vegetables are those where the male and female cellules are quite the same and advance to meet each other equally; thus possessing not only the same form, but also the same properties. In a small Alga bearing the name of *Ulothrix serrata*, the interior of certain cellules divides into two parts, which separate, then come together again and mingle anew into a minute mass which, when set at liberty, reproduces the plant entire. In other species the inside divides into small naked cellules, which are first set at liberty and for some time move briskly about in the water by means of cilia with which they are provided, before fusing into one another. These cellules are called zoöspores. The differentiation is further marked in certain species, the zoöspores of which have neither the same form nor the same properties. Some leave their positions to go to meet the others: these are the male cellules, the antherozoids; others make no movement at all and limit their *rôle* to that of waiting: these are the zoöspores. Similarly, in an Alga bearing the name of *Sphæroplea annulina*, there are found two kinds of filaments, brown and green. In the green filaments the protoplasm of certain cellules breaks up into a definite number of ovoid bodies which remain immobile; in the interim, the cellules of the brown filaments liberate mobile spores provided with two flagella: these spores, veritable male cellules, ply briskly about in the water and then proceed to fix themselves to the green filaments, the cellules of which are pierced by pores; through these orifices they penetrate into the cellules and fuse with the immobile ovoid bodies, which are nothing else than zoöspores.

The psychical phenomena attending this mode of conjugation may be still more complicated, as shown by the observation that Berthold has made upon the conjugation of the zoöspores of the *Ectocarpus siliculosus*. The *Ectocarpus* belongs to a group of algæ characterized by the presence of mobile spores which reproduce the plant. These zoöspores are little pear-shaped cellules, of which the tapering end is colorless, and the rounded end shows a brownish-green coloration, which is due to the presence of an extensive chromatophore; at the edge of the chromatophore a deep depression is sharply marked, which appears to be an eye. Every zoöspore is equipped, in addition, with two flagella, which rise from the same point of the lateral skirt of the anterior extremity of the body; one of these flagella points forwards and the other

Fig. 9. — Sexual reproduction of the *Ectocarpus siliculosus*. Different stages of the female zoöspore while entering the state of rest (after Berthold).

backwards. When the zoöspores are set at liberty and begin to swim about in the water, they pass each other by unnoticed. The female cellule does not draw about her the male cellules, from which, moreover, it differs by no morphological mark. But at a given moment the female zoöspore becomes distinguished from the male cellules by passing into a state of rest; whereto, the base of the anterior flagellum, which is laterally inserted, proceeds to blend with the anterior part of the body with the effect that the flagellum appears to rise from the extremity; during the same time, it

contracts, and presents at its free end a slight protuberance, which allows the zoöspore to fix itself upon
an immobile point; as to the rear flagellum, it slips
back upon the posterior part of the body which it
encompasses, and finally disappears.—When the female zoöspore has become motionless, the male zoöspores, hitherto indifferent, are seen to make towards
it and to surround it in a half-circle; the number of
zoöspores that thus meet, is quite considerable; it
frequently exceeds a hundred

(fig. 10). They let their second
flagellum float loosely behind
them, while they all direct their
anterior filament towards the female cellule; this filament they
draw back and forth over the
body of the female cellule; they
perform upon it real acts of feeling, the object of which is evi-

Fig. 10. — Sexual reproduction of the *Ectocarpus siliculosus.* Female zoöspore surrounded by male zoöspores.

dently to provoke in the female zoöspore a genital excitation, as what follows will prove. It happens at
times that several of the male zoöspores quit the
ranks and make off; they are immediately replaced by

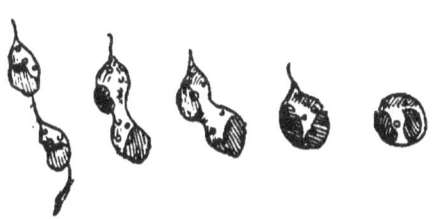

others who employ
their filaments in a
like manner, to stroke
the female. Finally,
upon the expiration
of a certain time,
one of the zoöspores
leaves the half-cir-
cle and approaches

Fig. 11. — Sexual reproduction of the *Ectocarpus siliculosus.* Successive stages of the
copulation of a female zoöspore with one of the
male zoöspores.

the female. The two zoöspores unite; after having
presented the series of changes marked in the figure,—

when the fusion is complete,—the female cellule loses its fixatory filament and the little zygote, the result of the fusion, is set free.

When the male zoöspore is obliged to go a long distance to reach the female zoöspore, it has been thought that the latter secretes a substance which acts upon the male cellule as a chemical excitant and which marks out for him the direction to follow. The supposition is quite probable; it was suggested by Strasburger, who had shown that the spermatozoids of the *Marchantia polymorpha* are attracted by the substance that issues from the archegonium. It will only be necessary to assist at an experiment of artificial fecundation with fish-spawn, in order to come to the same opinion. The sperm introduced into the liquid preparation does not spread about homogeneously in all directions; the spermatozoids are observed to whirl about the ovules in great masses; it must be supposed, further, that there is some excitation of an unknown nature which attracts the spermatozoid towards the micropyle, for this minute opening, of which the diameter is scarcely that of the head of a spermatozoid, is the only orifice through which the male element can enter into the ovule to fecundate it.

These ingenious opinions have been latterly confirmed by the very interesting experiments of M. Pfeffer, professor at the University of Tübingen, upon the movements of spermatozoids.* His investigations had to do principally with the spermatozoids of cryptograms. M. Pfeffer discovered that certain chemical substances have the property of attracting these spermatozoids.

---

* Pfeffer, *Untersuchungen aus dem botanischen Institut zu Tübingen*, Vol. I. Leipzig, 1884, p. 363.

The manner of conducting the experiment is as follows. A solution of the substance to be experimented upon is placed in small capillary tubes with a light-aperture of from five to seven hundredths of a millimeter wide. These capillary tubes dip into a watch-crystal covered with a liquid, wherein quantities of spermatozoids have been placed. Under these circumstances currents of diffusion are soon set up between the tube and the liquid in the watch-crystal, and when the substance experimented upon is the proper one, the spermatozoids are seen to follow the currents of diffusion and to penetrate into the tube.

·The substance exerting such attraction varies with the plants. The author began by experimenting upon the spermatozoids of certain ferns (*Adiantum cuneatum*). After a great many fruitless trials, one substance, and one only, proved to be effective: namely, a solution of malic acid or malate. It is to be presumed, then, that, in the organic kingdom, malic acid must be the substance acting as a chemical excitation upon the spermatozoids of ferns and guiding them towards the female cellule.

According to the hypothesis of Pfeffer, the actual process takes place in the following manner. The spore of a fern, falling upon humid ground, germinates and gives birth to a green cordate slip, the *prothallium*, upon which are developed the male organs or *antheridia*, and the female organs or *archegonia*. At a certain moment, elongate cellules, spirally twisted and extremely mobile, issue from the antheridium: these are the spermatozoids. They are equipped with vibratile cilia, by the help of which they are able to start in search of the female cellule. At the same instant, the female organ, the archegonium, opens and emits a mucilaginous substance, which must contain malic acid or a malate, for these compounds are the particular excitatory substance of the fern-spermatozoids. Thanks to a drop of dew that falls upon the prothallium, the spermatozoids swim around and·approach the female ovule, which attracts them by acting upon them with the malic acid.

A confirmation of this hypothesis is primarily the fact, that all substances tested, with the exception of malic acid and malates, remained completely inactive; another proof is, that malic acid is found in prothallium-decoctions of the *Pteris serrulata* and of the *Adiantum capillus veneris;* another proof still, is the circumstance that malic acid is largely diffused throughout the vegetable kingdom.

The author has made, in this connection, a series of very curious experiments upon the degree of concentration necessary to attract the spermatozoids. The lower limit at which attraction begins, is found in a solution containing malic acid in the proportion of one to 1000 parts. This the author has designated by a favorite word of the Germans: *Reiz-Schwelle*, or, in other words, the threshold of excitation.

When the solution in the watch-crystal contains one part malic acid to every thousand parts, in order to make the spermatozoids pass from the watch-crystal into the tube, the solution held in the tube must be thirty times as strong, or $30 \times 1 \cdot 1000 = 3 \cdot 100$. If the liquid in the watch-crystal contains one part malic acid to every hundred parts, similarly the solution of the tube must be thirty times as strong, that is to say three tenths.

The author justly compares the result of these experiments with the law laid down by Weber, which M.

Delbœuf has happily formulated as follows: " The slightest difference capable of being felt between two excitations of the same sort is due to an actual difference that increases proportionally with the excitations themselves." Thus, in order to tell that one weight is heavier than another, it must be heavier than the other by a fractional difference which varies from one third to one fifth according to the individual, be the original weight what it may. For example, to a weight of three grammes, in order that a difference may be made perceptible, we must add one third of three grammes or one gramme. To four grammes we must add one third of four grammes, or one and one third grammes, etc.*

According to Pfeffer, the application of Weber's law to his experiments is so exact that, when the solution of the tube is only twenty times stronger than that of the watch-crystal, the spermatozoids remain unaffected. Furthermore, the application of the law is not disturbed by changes of temperature varying within certain limits. Thus, down to a temperature of $+ 5°$ (41° Fahr.) the spermatozoids remain sensible to a concentration of liquid thirty times as strong as that in which they are.

Basing his calculations upon these experiments, the author has succeeded in determining the probable quantity of malic acid that must be contained in the archegonium. This quantity is probably in the proportion of three tenths.

The spermatozoids of the *Selaginella* are likewise wise attracted by malic acid and the malates. As regards the Marciliaceæ, the specific substance has not been discovered. The same failure, also, in the case

---

*Consult Ribot, *Psychologie allemande*, p. 161.

of the Hepaticæ. The author concludes from this, that the substance operating in these two cases can be little diffused throughout the vegetable kingdom. For the spermatozoids of the *Funaria hygrometrica* (Confervæ), the operative substance is cane-sugar. No other attracts them. The spermatozoids remain unaffected even by substances bearing the closest analogies with cane-sugar. We will cite, by way of example, fruit-sugar or levulose, grape-sugar or glucose, glycogen, manna, milk-sugar, etc.; these substances exert no attraction upon the movements of the spermatozoids, whereas cane-sugar exercises an attraction so powerful that the capillary tube becomes at once crammed with them. The excitation first induces in the spermatozoid a movement of direction: the body is brought into a position enabling it to reach the tube by movement in a straight line. The same phenomenon has been observed by Strasburger in the case of Algæ zoöspores; when these minute beings are attracted by a chemical or luminous excitation, the first thing that happens is the directing of the body towards the attracting source.

A solution of one in one thousand parts is sufficiently concentrated to draw the spermatozoids of Mosses into the capillary tubes. The "threshold of excitation" for them, accordingly, is the same as for the spermatozoids of ferns. Furthermore, Weber's law is in this instance again verified; only, in order to have the chemical excitation produce a different attraction, it must be stronger than the first in the proportion of 50 to 100. In the experiments upon the spermatozoids of ferns the ratio is a little smaller; being only 30 to 100.

The question presented itself to the author as to

whether, by increasing the degree of concentration, a point would not be reached where attraction would change to repulsion; he has not made the experiment, but he has noticed that great numbers of spermatozoids still penetrate into the tube when containing a solution in the proportion of 15 to 100, notwithstanding the fact that they there meet a speedy death.

The general conclusion to be derived from these numerous experiments is, first, that the spermatozoids are sensible to certain chemical excitations, and consequently, that in every group of plants there exists a special substance acting the part of a specific excitant towards the spermatozoids. The author does not hesitate to regard the spermatozoids as a physiological re-agent of such substances, allowing feeble traces of the same to be detected in a liquid solution. He thus comes to form a *spermatozoid test*, which is not without analogy with the *Bacteria test*, invented by Engelmann. An application of the test is the following: a decoction of herbs having presented the property of attracting the spermatozoids of Mosses, the author concluded that the decoction must contain canesugar.

VIII.

THE PHYSIOLOGICAL FUNCTION OF THE NUCLEUS.

It would be of the highest importance to know what is the seat of the phenomena of the life of relation in the bodies of Micro-organisms. We have seen that Micro-organisms are the equivalent of a simple cellule, composed, according to the classic plan, of a protoplasm, of a cellular nucleus, and of an enveloping membrane.

Each of these elements plays a part of special importance in the vital phenomena of these beings. Long since, scientists have attributed movement, sensibility, and the prehension of foods, to the protoplasm. This was the result of direct observation. While observing an Amœba, for example, the protoplasm is seen to undergo modifications of form and to throw out pseudopods, either for the purpose of effecting a change of position, or to seize alimentary substances. The protoplasm, accordingly, seems to be the sole agent of all these phenomena. Likewise, the vibratile cilia of the Ciliates, which are at once organs of motion, prehension, and touch; the suckers of the Acinetinidæ, which are special organs of. prehension, are nothing else than outward expansions of the protoplasm proper.

As regards the enveloping membrane, the same cannot discharge any psychical function: firstly, because it is a product of protoplasmic secretion; and, secondly, because it is wanting in many Protozoans and even in many animalcula≈quite high in point of organization that, despite their nudity, exhibit marks of psychic life just as complex as those observed in Infusoria having a cuticle. The part acted by the nucleus does not so clearly manifest itself to direct observation; it executes no movements in the ordinary conditions of life; it remains motionless in the centre of the animal's body, surrounded on all sides by the protoplasm; unlike the latter, it is not in direct contact with the outside world.

The first phenomena that have enabled us to conjecture as to the significance of the nucleus, have to do with the division of cellules; when a cellule divides, the nucleus comes into action, it exhibits certain movements, and passes through complicated stages which have been given the name of caryokinesis.*

But these complex phenomena simply show the function of the nucleus as an histological element; they do not afford any disclosures as to the physiological *rôle* of the nucleus in the cellule.

Other observations have enabled naturalists to surmise what phenomena are subject to the action of the nucleus. In 1881, Balbiani called attention to individuals, belonging to the species *Paramæcium aurelia*, that were destitute of a nucleus and which nevertheless possessed the power of locomotion the same as ordinary individuals; whence, he concluded that the nuclei exerted no influence upon the phenomena of individual life. Shortly afterwards, Gruber observed small specimens of the *Actinophrys sol* which absorbed nutriment, changed their position in the liquid, and even fused with each other (zygosis), but which were nevertheless destitute of a nucleus.†

The idea then occurred to Gruber, and to Nussbaum likewise, to divide the Micro-organisms by artificial means into several fragments, of which some would contain a nucleus and others not, and then to watch what would come of it. Gruber, to whose experiments the most importance attaches, chose as his subject of trial the *Stentor cœruleus*, a ciliated Infusory of great size, which exhibits a nucleus resembling a chaplet of beads (moniliform). He afterwards continued his experiments upon other species, and his conclusion was, that the power to regenerate lost parts belonged to all Protozoans, but that this phenomenon only took place when the isolated fragment contained

---

* κάρυον, the nut, and κίνησις, motion, disturbance.

† Contributions to the *Biologisches Centralblatt*, 1885. p. 73.

some portion of the nucleus; in which case the animal reproduces all the organs it has lost in consequence of its dissection. Furthermore, the process of the formation is exactly the same as in the spontaneous division of these same Infusoria. The excitation caused by their removal is accordingly of the same character as the unknown excitation that provokes the natural division of the body.

From these experiments, the part acted by the nucleus is indicated by complete evidence. Gruber shows that in a single instance only can a fragment without a nucleus form itself anew; and that is, when the fragment contains an organ in course of formation, as happens, for example, during the spontaneous division of the animal. This amounts to saying, that the presence of a nucleus is necessary to give the impulse to the formation of the organ, but that it is not necessary to the completion of the organ when the impulse has once been given.

Lastly, if the fragment is totally destitute of a nucleus, it does not re-form itself so as to constitute a complete animal again; if the fragment possesses neither mouth nor peristome, it does not reproduce a new mouth and a new peristome; yet the fragments continue to live and to move. The absence of a nucleus does not suspend the functions of motion, sensibility, nutrition, or growth. This conclusion is, in our estimation, too sweeping, as we shall see further on.*

M. Balbiani has recently repeated these experiments of artificial division, and, while confirming in

---

* We have taken as our guide, with the permission of M. Balbiani, the lectures delivered by that eminent authority at the Collège de France, in May, 1887.

general the results of Gruber as to the function of the
nucleus in the vital phenomena of ciliated Infusoria,
he has endeavored to fix with more exactness a certain number of important points.   His first experiments, like those of Gruber, were conducted upon the
Stentor cœruleus, a species of which the size renders it
better adapted to this sort of experimenting.   In an
observation which we shall take as a type, and which
is represented by the figure sent to us by M. Balbiani,
the body of the Stentor is cut by two transverse sec-
tions; three divisions are obtained, each of which contains a fragment of the nucleus. We will remember
that the nucleus of the Stentor is like a long string of
beads; it is not at all rare to see a fragment of a Stentor contain one or more beads.

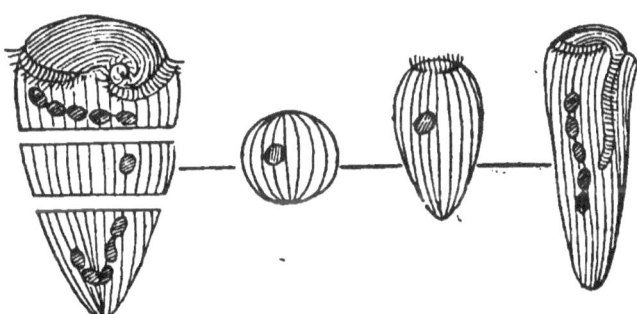

Fig. 12.—Artificial division of the *Stentor cœruleus.*
(After Balbiani.)

Let us follow the phenomena presented in the
middle segment. This segment contains only a single
grain of the nuclear chaplet; directly after severance,
it assumed a globular shape; the day following, it had
lengthened, had grown a tail at the posterior extremity, and upon the anterior part there had appeared,
distinctly outlined, a crown of cilia longer than those
upon the body; in other words, a peristome had

formed; the day after, the fragment had increased considerably in bulk, and in two days more the animal had formed a mouth. During this time, the nuclear grain had multiplied: five, in fact, were counted. The animal had the normal form; its size, however, was a little smaller than that of the ordinary Stentors. Thus, through the action of a small quantity of nuclear substance, the fragment has been completely reconstructed.   .

It frequently happens that the artificial severing of the animal causes various deformations in the fragments. The deformation disappears with the greatest rapidity in fragments containing nuclear substance. The wound heals instantly; directly after severance, the two edges of the wound are seen to adjust themselves to each other.

In all these particulars, the experiments confirm the results obtained by Gruber.

M. Balbiani desired to ascertain what would happen if the division were made during the state of conjugation.

Conjugation, as we know, aims at replacing an old, spent element, that has lost its physiological properties, by an element of new formation proceeding from an attendant nucleus (nucleolus) exchanged between the individuals in conjugation. The point in question was to ascertain whether the nucleus that was beginning to disappear, had lost its regenerative power. In the Stentors, during conjugation, this old nucleus breaks, and its nuclear globules are scattered to all parts of the protoplasm. When at this stage, the body of one of the Stentors is divided in such a manner that the fragment contains some of the scattered globules that came from the old nucleus. It is quite evident that such a fragment is obtainable only by mere accident.

In an experiment which we again cite as a type of many others, the fragment containing the elements of the old nucleus tends to reconstruct itself; this fragment, which represented the posterior part of the animal, presented, the day following, a rudimentary peristome; the reconstruction did not go beyond this point: it was left incomplete. Accordingly, the old nucleus loses its power of regeneration.

As to the phenomena presented in fragments containing no nuclear substance, M. Balbiani has made decided advances in the question; he has completed the experiments of Gruber, he has also corrected them, and he has reached conclusions essentially different.

In order to understand more thoroughly the phenomena connected with the absence of nuclear substance, the author has directed his attention to another species, the *Cyrtostomum leucas*, which has the advantage that it can be kept longer alive than the Stentor can, on a glass slide holding a drop of water. The Cyrtostomum is a large ciliated Infusory of more than four-tenths of a millimeter in length. Its protoplasm is differentiated into two layers, one of which, the cortical, encloses very heavy trichocysts; the other, the endoplasm, holds alimentary substances. The animal exhibits upon one of its faces a mouth, shaped like a long narrow buttonhole, and upon the other face a contractile vesicle, from which crooked and anastomosed passages radiate. It is easy, by making a transversal division, to obtain fragments without nuclear matter; the nucleus of the Cyrtostomum being formed of a single, round mass. But it is not easy, on the other hand, to obtain fragments likely to live, since this animalcule has a dense ectoplasm, and, when severed, this layer, which is not very retractile, does not grow together again and close the wound; the sides remain separated, the water comes in contact with the endoplasm, which swells, bulges out, and runs from the wound; the animal may thus void itself completely, dying of diffluence. It occasionally happens that the animal voids itself only in part, and that the nucleus escapes with a small piece of the protoplasm. Then, if the wound draws together, we get a fragment that has thrown out the nucleus by its own action.

We shall not speak of the actions of the fragment containing nuclear substance; they are the same as observed in the case of Stentors: the fragment rapidly reconstructs itself and re-forms a complete animal.

Let us mark more closely the fragment without nucleus. Such fragments continue to live for some time; they have been kept alive as long as eight days; but they do not reconstruct themselves; they do not even assume a regular form; the part of the body facing the section retains its obliquity of truncation. At the start, for the first few days, the movements continue; a curious circumstance connected therewith is, that the fragments continue to move in the direction in which they would have moved if they were placed together to form a complete individual. The vibratile cilia are in no wise altered; they shake with the same animation as before. Only the movements of the animal are a trifle irregular; but they exhibit the same marks of volition as seen in normal individuals. The vesicle continues to contract.

The power to seize food is also retained when the

fragment without nucleus contains the mouth; the mouth ingests alimentary substances. If the Cyrtostomum be given grains of potato fecula, which it is very partial to, the fragment without nucleus, but with a mouth, swallows these grains and fills itself with them. It is not known whether it digests them.

This much was observed in the first stages, and Gruber was wrong in stopping at this point.

At the expiration of a certain time, varying between the third and fourth day, alterations of structure are noticed in the fragment that are probably traceable to the absence of the nucleus. One of the first to take place is the disappearance of the marks of differentiation which we have observed to distinguish the endoplasm from the ectoplasm. The dark granules that fill the interior of the body congregate in the centre ' by abandoning the peripheral part; then these granules scatter and come to a position just beneath the cuticle, which denotes a deliquescence of the plasma. The layer of trichocysts undergoes changes and disappears. All these alterations result from an actual disorganization of the plasma. The contractile vesicle shrinks, its pulsations decrease, the radiating passages disappear. The body of fhe animal, which in its normal condition is elongate, becomes rounded; its movements flag and consist of nothing but a rotation of the body about its own axis; at last the animal becomes motionless and dies of diffluence.

These changes are not due to lack of sustenance, as one might suppose; for fragments that have a mouth and swallow food, pass through the same alterations as those that have no mouth.*

---

* M. Balbiani has informed us, upon request, that the fragments of Cyrtostomum furnished with nucleus can be kept alive for a much longer time

It is superfluous to insist upon the importance of these results, obtained by a method that might be called experimental physiology applied to unicellular organisms. Although the experiments have been made solely with ciliated Infusoria, the results of the same may be extended to all cellules, for the Infusoria are nothing more than autonomous cellules living an independent life.

The conclusion from the above researches of M. Balbiani, which, as we have seen, go far beyond those of Gruber, is, that the nucleus is not necessary merely to the regeneration of the parts, as the German professor believed. The error made by Gruber arose from the fact that he did not follow the career of the fragments deprived of a nucleus long enough; if he had continued his observations, he would have seen that the fragment becomes gradually disorganized. The nucleus, accordingly, has not merely a formative power; it does not merely regulate alimentation, re-adjustment of form, and the healing of wounds; it has not merely a regenerative power, enabling the plasma to reconstruct complete the organs lost by artificial severance. The nucleus is, besides all this, an essential factor of the plasm's vitality. If a fragment of protoplasm be deprived of its nucleus, the fragment remains alive for some time, but afterwards undergoes disorganization.

Such are the facts, extremely complex, and consequently difficult to summarize by a formulated statement.

---

under the same conditions (that is, in a drop of water on a glass slide kept in the moist chamber of Malassez): in this way it is possible to keep them alive for the space of a month, by introducing into the liquid a few Infusoria to serve them as food. On the other hand, the fragments deprived of nucleus by section live for only eight days at the most.

We certainly cannot regard the protoplasm as inert matter; but what appears probable is, that the protoplasm receives from the nucleus the communication, the delegation of physiological powers. The nucleus is in a certain sense the focal seat of life in all its forms.

If we get rid of the nucleus by artificial section, the fragment of enucleated protoplasm continues to live for some time, having received from the nucleus an impulsion that has not yet been exhausted; but after a certain length of time, the impulsion given by the nucleus not being renewed, the protoplasm runs its course and dies.

From the psychical point of view, which more particularly occupies our attention here, how are the results of these experiments in cellular vivisection to be explained? When a fragment of an organism, deprived of nuclear substance, is seen to move about freely and with the same activity as if it still possessed its nucleus, we are constrained to admit that the phenomena of the life of relation, or movement and sensibility, have their seat in the protoplasm. But it is probable that such physiological capacities as the powers of nutrition, are not inherent in protoplasm; they depend immediately upon the presence of the nucleus, for they disappear little by little and finally vanish a few days after the removal of the nucleus.*

It may be mentioned in passing, that there are certain psychical properties which the nucleus apparently does not transmit to the protoplasm, but which it retains for itself; this is the case with the instinct of gen-

---

* The difficult question here, is to ascertain whether the psychical properties of the protoplasm are destroyed through the direct effect of the disorganization of the plasma, or whether they disappear a short time before the process of disorganization and in consequence of the absence of nuclear substance.

eration. We have already seen that, during the epidemic periods of conjugation, the Paramecia which have their nuclei overrun with parasites cease to conjugate with animals of the same species. The destruction of their nucleus by the Bacteria produces in the Paramecia the effect of actual castration.

The removal of the nucleus, accordingly, causes the interruption of the following functions and in the following order as to time:

1. The regenerative and reproductive property of the plasma;

2. The vitality of the plasma, and the psychical functions.

The psychologist will notice with interest that the psychical function of the protoplasm outlives the regenerative function for an appreciable length of time; a fragment of a cellule which, having been mutilated by the act of severance, is unable to correct its outward form, or to secrete a fresh cuticle, or to reconstruct its lost organs, is nevertheless still capable of perceiving sensations and of responding thereto by movements. Psychical life is consequently a property of living matter which appears to be less complex than the regenerative property, inasmuch as it ceases later.

To summarize, the nucleus plays the primordial *rôle* in the cellule; if, to use an old comparison of Aristotle's, we compare the protoplasm to the clay, we must compare the nucleus to the potter that fashions it. The nucleus comprehends all the physiological properties, the totality of which goes to constitute life.

It is interesting to note what perfect accord prevails between these recently discovered facts and the phenomena relative to fecundation. Fecundation consists in the fusion of two nuclei, of which one proceeds from the male, and one from the female. Thus, it is through the intermediary office of the nucleus that all the faculties, all the properties possessed by the parents,—the form of their bodies as well as their psychical faculties,—are transmitted to the embryo; as we have just remarked, therefore, all these properties must be comprehended in the nucleus, in order to pass into the embryo.

We must note further, that the embryo takes from the mother something besides the nucleus. While it is connected with the father through the head of the spermatozoid, which has the morphological value of a nucleus, it receives from the mother not only the female nucleus but also the vitelline plasma of the ovule; now, as the embryo does not exhibit a greater morphological likeness to the mother than to the father, we may thence infer that the vitelline protoplasm inherited from the mother exerts no formative influence upon the development of its body.

These are not the only facts the connection of which we desire to show with the results of experiments upon the function of the nucleus. It will be well to point out here, how reproduction is effected among organisms which, besides their nucleus, possess other differentiated organs. The best known and perhaps the most general mode of reproduction is fissiparity, which consists in a division of the entire body into two equal parts. If we closely follow the course of this phenomenon in any organism whatever, we shall find that the division begins by a multiplication of the principal organs of the body. The nucleus begins by lengthening out and assuming a position perpendicu-

lar to the plane of division. The first organ that multiplies is the flagellum; it does not split into two parts,
as several English authors have supposed; according
to the observations of Bütschli and of Klebs, a second
flagellum is formed complete. The pigmentary spot
also does not divide into two parts; the old eye remains by a sort of preference with one of the parts,
while the other part acquires a new eye, formed complete; this is likewise the case with the mouth and the
œsophagus. There are only two elements that multiply by division: the chromatophores and the nucleus
Now, when we note that the chromatophores contain a
body, the pyrenoid, which exhibits the closest analogy
of chemical composition with the nucleus, we may
properly say that the nuclear elements of the cellule
are the only ones that do not reproduce by neoforma-
tion at the expense of the protoplasm, as is the case
with the cilia and the flagella.

· The reason for this mode of multiplication by nuclear elements will be comprehended, if we consider
the matter in the light of experiments made upon the
formative properties of the nucleus. We have seen,
in fact, that the nucleus can regenerate the protoplasm,
but that the protoplasm cannot regenerate the nucleus.
We now see that the regeneration of organs lost in
consequence of the spontaneous division of cellules, is
subjected to the same law as the regeneration following upon artificial division; the protoplasm cannot regenerate a nuclear element any more in the one case
than in the other; in order to effect reproduction,
therefore, this element must divide.

IX.

CONCLUSION.

THE conclusions relative to psychological phenomena arrived at in the foregoing treatise, are in contradiction with the opinions generally received upon the psychology of the cell. Scientists have held, that cellular psychology is represented wholly and solely by the laws of irritability. In his *Essai de Psychologie Générale*, a work in so many respects remarkable, M. Richet has assumed the advocacy of this view; the correctness of which we have no hesitation in disputing. In the work just mentioned, the distinguished professor has written the following:

" There exist simple beings which appear to be nothing more than a homogeneous assemblage of irritable cellules. Motory reaction, consequent upon irritation from without, constitutes their life of relation. Irritability is their life complete, but this, in effect, is psychic life; so that cellular irritability can be considered the same as *elementary psychic life.*"

From an attentive perusal of this passage it will be seen that M. Richet brings within the category of irritability, not only unicellular organisms, but also pluricellular organisms formed by the union of homogeneous cellules.

M. Romanes, in his work upon *Mental Evolution*, without coming to a conclusion so definite as M. Richet, seems to us to have reduced the psychic activity of proto-organisms to within very narrow limits. We are impressed with the fact upon glancing over his *Diagram of Mental Evolution:* he recognizes nothing but excitability, for example, in the ovule and spermatozoid of man. This is manifestly erroneous.

The sexual elements, and especially the spermatozoid, of all unicellular organisms are certainly the ones which show the most highly developed psychical functions: the act of seeking and approaching the ovule, which is frequently situated at quite some distance from where the male element is deposited; the length of road to be traveled; the obstacles to be overcome; all point to faculties in the spermatozoid that are not explainable by simple irritability.

Hitherto, apparently, writers who have essayed to present the psychology of micro-organisms, have contented themselves with schematic notions instead of basing their theories upon the direct observation of these interesting creatures. By the aid of exact data, we have shown that in both vegetable and animal micro-organisms phenomena are encountered which pertain to a highly complex psychology, and which appear quite out of proportion to the minute mass that serves them as a substratum.

We shall first of all advert to the term irritability, which, though long in use, has not in our opinion been happily chosen; since it is in the highest degree ambiguous, and not suggestive of an exact signification. We might call to mind in this connection, the reflection made by Kant upon obscure properties, which he compares to easy-chairs upon which the mind unbends itself and rests. In place of discussing words, let us endeavor to discuss facts.

What are we to understand by irritability? We may give the expression a very broad or a very restricted meaning. We may make it express the property which every organism possesses, of reacting upon excitation. In this general sense we may say that irritability includes within itself all of psychology, the most highly developed, as well as the most elementary; for upon ultimate analysis every psychical manifestation consists in a response to an excitation.

Evidently, it is not in this general and somewhat common sense that M. Richet has intended to employ the word. For a more exact definition, let us consult his work, of which, a whole chapter, the first, is devoted to this subject; the author enumerates and develops at length the laws of irritability:

1. Every action that modifies the actual condition of a cell is an irritant of that cell.

2. Every external force, provided it has a certain intensity, is capable of inducing cellular irritability.

3. The movement in response to irritation is proportional to the excitation.

4. The movement in response to irritation is, for equal irritations, stronger in proportion as the equilibrium of the cell is less stable; in other words, stronger in proportion as the cell is more excitable.

5. The response to the irritation, is a movement in the form of a wave, which has a very short latent period, a period of ascent, correspondingly brief, and a very long period of descent.

6. The movement of the cell upon irritation is, for equal irritations, stronger in proportion as the irritation has been more sudden.

7. The movement in response to a brief irritation lasts much longer than the irritation has lasted.

8. Forces which, alone, appear impotent, become effective when repeated; for they have, in spite of their apparent inefficacy, increased the excitability of the organism.

The statement of these various laws, gives the term irritability a precision which it lacked. M. Richet had in view particularly the muscular fibres, and the laws of irritability are only supposed to cover a series of physiological experiments made upon the reaction of a striated muscle. They are not, then, hypothetical laws, but are much rather particular experiments generalized and extended to undifferentiated protoplasm. It is proper to remark here, that we have not as yet been able, by means of direct experiments, to ascertain from life the laws of irritability in undifferentiated protoplasm. The experiments made upon this point,— for instance, the experiment causing the protoplasm of a detached cell to contract by means of an electric current,—have not yet been brought to a precise result; for the structure of protoplasm is so delicate and so complex, that even the slightest excitation suffices to produce an alteration, and since it is difficult to distinguish the contraction of the protoplasm from its coagulation. But we shall pass by this subordinate question.

The question now remains, whether the complicated experiments made in muscular physiology, which M. Richet generalizes and extends to the physiology of all cellules, include and comprehend the whole psychology of an independent organism, and whether we may say with M. Richet, that irritability (thus understood) represents all of cellular psychology.

Plainly not. The numerous facts which we have cited in the foregoing essay, transcend the too narrow limits within which it has been attempted to confine the psychology of the cell. We shall restrict ourselves to the mention of one of these phenomena, to show the complexity of the psychic life of micro-organisms: it is the existence of a power of selection, exercised either in the search for food, or in the manœuvres attending conjugation. This act of selection is a capital phenomenon; we may take it as the character·istic feature of functions pertaining to the nervous system. As Romanes has indeed observed, the power of choice may be regarded as the criterion of psychical faculties. Going farther, we might be able to say that selection comprehends the properties of the nervous cellule, as irritability comprehends the properties of the muscular cellule.

Scientists have endeavored to explain the mechanism of this choice. They have pretended to solve it by saying that it was dependent upon the relation between the chemical composition of the cellule making the choice and the chemical composition of the body selected.

Such explanations are purely verbal. Undoubtedly, the faculty of selection, of which protoplasm seems to be possessed, is founded in the character of its chemical composition. Chemistry lies at the basis of physiology, but chemistry does not explain physiology, and it is quite evident that that property which protoplasm possesses of making a choice between several excitations, is a physiological property.

However that may be, we may resume all the foregoing into the statement that every micro-organism has a psychic life, the complexity of which transcends the limits of cellular irritability, from the fact that every micro-organism possesses a faculty of selection; it chooses its food, as it likewise chooses the animal with which it copulates.

M. Richet has defended his opinion in opposition to the one I have propounded, in a note published in the *Revue Philosophique* for Febuary 1888, wherein he speaks as follows:

At the beginning of his essay upon the Psychic Life of Micro-Organisms (*Revue Philosophique*), M. Binet expresses himself as follows: "In the lower beings that represent the simplest forms of life, we find manifestations of an intelligence which greatly transcends the phenomena of cellular irritability. Thus even on the very lowest rounds of the ladder of life, psychic manifestations are very much more complex than is usually believed, and the conception of cellular psychology which some very recent authors have formed, seems to me a very crude analysis of the most delicate of phenomena."

As I have upheld in my *Essai de Psychologie Générale*, and in some measure—however little—developed this admitedly old idea, that cellular irritability is the beginning of psychical activity, I request the permission to speak in defence of an opinion so roughly handled by M. Binet.

Now, it appears to me that M. Binet has allowed himself to become involved in illusion respecting the word cellule. A cell, in the eyes of the embryologist and the morphologist, has a well-defined meaning. But M. Binet does not seem to have comprehended the fact, that for the physiologist and the psychologist, the essential condition of cellular unity is homogeneity. It is possible that the infusoria, the strange story of whose life M. Binet relates to us, are single-celled organisms. I am in no wise qualified to decide as to this; but whether a single cell, or a group of cells, it matters little, in my opinion, provided the single cell is differentiated to the same degree that it would be if composed of several cells not homogeneous.

I appeal to M. Binet himself and to the cuts of his essay. When he shows us an *Euglena* with eyes, æsophagus, mouth, contractile vesicle, contractile reservoir (fig. 6); when he carefully describes the shape of the flagellum, the nettle-like tentacles, the tongue-shaped organs, the ocular spots, the trichocysts, and the peristome; when he assumes special *nervous centres* endowed with various attributes (p. 22): he cannot induce us to admit that the psychology of these complicated organisms is the same as the psychology of the simple cell. I repeat, it is quite immaterial to me that people affirm by the authority of embryology that this or that is a single cell. If that cellule have ocular organs, a nervous system, a mouth, an æsophagus, and a heart, I shall, despite any and every hypothesis of the embryologists, refuse to regard it as being physiologically a homogeneous cell, as is, for example, a muscular fibre.

The size will not affect the matter at all. The same desires, says Montaigne, stir mite and elephant alike. The psychic life of the bee is as complex as that of the whale, and if a microscopic infusory possess eyes, mouth, prickles, and heart, it evidently possesses them in order that it may make use of them, and accordingly I shall treat it as a complex organism upon the same ground that I do a snail or a grasshopper. Embryology will not force me to the extremity of regarding such a creature as a simple organism because it is derived from a single cell.

In my opinion, therefore, it is that unfortunate word *unicellular*, that has made M. Binet believe that, Infusoria being unicellular organisms, the elementary psychology of the cellule applied to them. M. Binet has allowed himself to be deceived by a word— a thing that often happens in matters of science. For my own part, in order to avoid any confusion, I would like to say that the elementary psychology of the cellule ought not by rights to be applied to anything except to homogeneous cellules; for the psychology that has to do with complex cells—real organisms with organs and apparatus of their own—must certainly be as complex as the psychology of animals wholly differentiated.

The laws of irritability act in all their simplicity and rigor among simple beings. In fact, in every instance of investigation into the nature of simple organisms, or such as appear simple by the optical instruments at our disposal (a fact that does not always rigorously prove their simplicity), as bacteria, for example,—we find that chemical irritability is the apparently sole law of movement. What else, indeed, are the movements of those bacteria so thoroughly studied by M. Engelmann, if not an affinity for oxygen, in other words the simplest and most universal chemical phenomenon in all nature?

And so the *critique* of M. Binet will not stand. On the contrary, it seems to be well established that complex organisms, whether single-celled or many-celled, have a psychology corresponding in *complexity* to the degree of differentiation their organs have attained, while simple beings—and they are simple only if homogeneous—have a *simple* psychology which is probably contained in the laws of Irritability only.   CH. RICHET.

My reply to the letter of M. Richet, published in the same number of the *Revue Philosophique,* may be offered as a general conclusion to my work. With the

omission of all polemical features, it is in substance ·
as follows:

In giving the psychology of these microscopic creatures the
name of cellular psychology, I have not invented a new term, nor
given a new sense to an old one.   Quite some time before me, M.
Hæckel had made a study of cellular psychology and his investigations, like my own, were based entirely upon the observation of
animal and vegetable micro-organisms.   Furthermore, micro-organisms being represented by a single cellule (and this doctrine is
.now universally accepted), the study of their psychical manifestations can, in my opinion, with perfect propriety be styled cellular
psychology.

M. Richet takes exception to the use of the latter expression;
, but he does so while substituting for the old definition of the word
cell, one quite his own.   To him, a micro-organism like the Euglena, which has an eye, a mouth, an æsophagus, and a contractile
vesicle, would not be a cellule.   To admit the latter view, means,
in his own words, to become involved in illusion respecting the
word cellule.   In our judgment, the question here is by no means
one of optical illusion, but one of verbal definition.   What, accordingly, is a cellule?   " For the physiologist and psychologist,"
says M. Richet, " the cellule has not a distinct entity, or, at least,
that entity, that unity, lacks an essential condition, namely, homogeneity."

To M. Richet, the cellule is a homogeneous body; a body that
comprises differentiated parts is not a cellule.

It is unnecessary to remark upon how far the latter conception of a cellule diverges from the usual and commonly accepted
definition of the word.   Hitherto, scientists have understood by
the term cellule, a body made up of the union of two essential
parts, a quantity of protoplasm and a nucleus. The scientific world
argues as to whether elementary forms exist which do not contain
a nucleus and which should be termed *cytodes*, as proposed by M.
Hæckel.   The careful observation of micro-organisms by means of
perfected technical processes has enabled us to discover hundreds
of nuclei in the very cellules which M. Hæckel classed among the
cytodes.   Such is notably the case with many algæ and lower-class
fungi.   The Moners—a group of micro-organisms believed to have
no nucleus—grow numerically less and less, in proportion as they
are more carefully studied.   It is true, we are now no more able

than formerly, to show the presence of a nucleus in bacteria; but that does not prove that the bacteria have none. Our knowledge of the morphology of microscopic organisms is wholly relative, and depends upon the degree of perfection attained by technical science. When we bear in mind that the presence of a nucleus remained for a long time unobserved in organisms several hundred times larger than the bacteria, we ought not to be surprised at having been unable to discover one in these smaller creatures.

We may even go further, and question the material existence of a body formed solely of protoplasm, basing our opinion upon the experiments of Gruber, Nussbaum, and Balbiani, as reported in my article, and upon the more recent observations of Klebs which are in perfect agreement with the results of the investigators just cited. All have shown that the nucleus is an element essential to the life of the cellule, and that, when a fragment of a cellular body stripped of a nucleus is procured by artificial section, this fragment does not reproduce the organs it lost by being severed; it does not heal its wound, it does not refashion its form, and, what is more, at the end of a certain time its protoplasm, being withdrawn from the influence of the nucleus, suffers complete disorganization. These experiments were made not only upon animal micro-organisms, but upon vegetable cells also. They prove the primordial importance of the nucleus in the cellule and thereby render doubtful the existence of cellules deprived of a nucleus.

Since every cellule contains, in all likelihood, two distinct differentiated elements, the protoplasm and the nucleus, which have neither the same physical structure, nor the same chemical nature, nor the same physiological functions, we may understand that it would be exceedingly difficult to name a single instance of a simple homogeneous cellule. It is the proper place to add that neither protoplasm nor nucleus, each regarded by itself, are homogeneous substances. It is unnecessary to enumerate all the investigations that have been made upon this point. Let us call to mind merely the fact that from the morphological point of view protoplasm appears to be composed of two substances, a homogeneous semi-liquid substance and a firmer substance exhibiting, as authorities upon the subject say, sometimes the form of detached filaments and at others a structure of a reticulate character.

At the present day, accordingly, it is impossible to allow that homogeneous cellules exist, without falling back to Dujardin's theory of the sarcode.   There are really no simple organisms, and such as appear so are merely imperfectly known.

However, it will not do perhaps to take literally the terms employed by M. Richet.   When he speaks of homogeneous cellules it is possible that he wishes to speak merely of cellules in which, aside from the nucleus, no other differentiated organ is to be found.

Now, it is quite important to note that, even of organisms made up simply of protoplasm and nucleus, the psychology is extremely complicated, and is not contained exclusively, in the laws of irritability.

The *Vampyrella Spirogyræ* classed by Zopf among the animal-fungi, and the place of which is still so little known, is a being the body of which is composed of protoplasm and nucleus simply. So far no other differentiated organ has been found in this creature, except from one to four contractile vesicles.   Employing the terminology of M. Richet, perhaps we ought to call this being a simple cellule; yet this simple cellule has quite a complicated psychology: it exercises choice in the selection of its food, attacking Spirogyra only.

The same is the case with the *Monas amyli*, which, having neither eye nor mouth, represents to M. Richet a simple cellule; still, the *Monas amyli* exercises choice in selecting its food, as it feeds exclusively on grains of starch.

The structural elements of the tissues do not differ from the micro-organisms whose psychological history I have endeavored to unfold, so much as might be imagined: they show the same powers of selection, and on this point I shall only instance the epithelial cellules of the intestines or the phagocyte cellules, the attributes of which I have described in my essay, and which are able to discriminate, for instance, between bits of fat and particles of coal, absorbing the former and leaving the latter.

I repeat it, therefore, no living cellule, strictly so defined, is a simple cellule, and I do not think that M. Richet has advanced a fitting illustration in mentioning the muscular cell, for the latter is one of the most highly differentiated that there are.

I cannot imagine, accordingly, to what elements, to what beings clearly defined, we could apply the simple-cellular psychology reduced to mere irritability, that M. Richet asks me to distinguish from the complex-cellular psychology, which would be exclusively reserved for the animal and vegetable micro-organisms that I have described.

It appears to me that this simple-cellular psychology lacks a foundation; it is a conception of the mind, rather than a study based upon observed facts.

In M. Richet's book I find no indication as to what sort of beings he means to distinguish thereby. He contents himself (pp. 20 and 27) with speaking of simple beings without otherwise defining them. Towards the close of his remarks upon my work, M. Richet cites an instance of simple beings, viz., the bacteria; in his judgment, chemical irritability appears to be the sole law conditioning their movements. What are the movements of the bacteria, he asks, if not an affinity for oxygen; in other words, the simplest and most universal chemical phenomenon that exists in all nature?

In our judgment the latter phrase is to be taken metaphorically. We believe that as yet no one has demonstrated that the movements of a living being, in moving towards a distant object, however simple they may be, can be explained merely by a chemical affinity acting between that being and that object. It is certainly not chemical affinity that is acting, but much rather a physiological need.

Psychic life, like its substratum, living matter, is, when closely studied, an exceedingly complex subject. This fact is, with me, a profound conviction ; it rests, not upon abstract ideas and methods, but upon the observations that I have given, observations that are not founded upon my own personal authority alone, but which are drawn from the highest authorities, and most of which I have been able to verify with my own eyes.

ALFRED BINET.

APPENDIX.

The subjoined cuts, explanatory of the conjugation of Micro-organisms, refer respectively to the descriptions on pages 67, 67 and 68, 68, 69 and 70, 71 and 72.

Fig. 7 *a.*—Positions preliminary to conjugation among the *Paramæcium aurelia.* (Balbiani.)

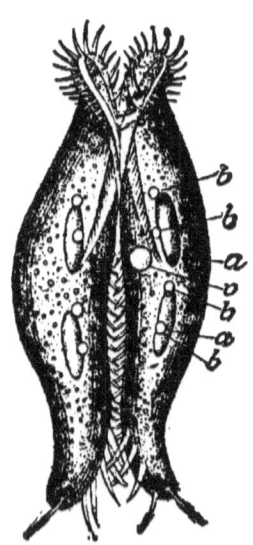

Fig. 7 *b.*—Several pairs of *Stentor cæruleus* fixed upon a conferva filament, enlarged fifteen diameters (after Balbiani).

Fig. 7 *c.* Position preliminary to the copulation of the *Stylonychia mytilus.* The two animals are superimposed upon each other by their ventral faces (Balbiani).

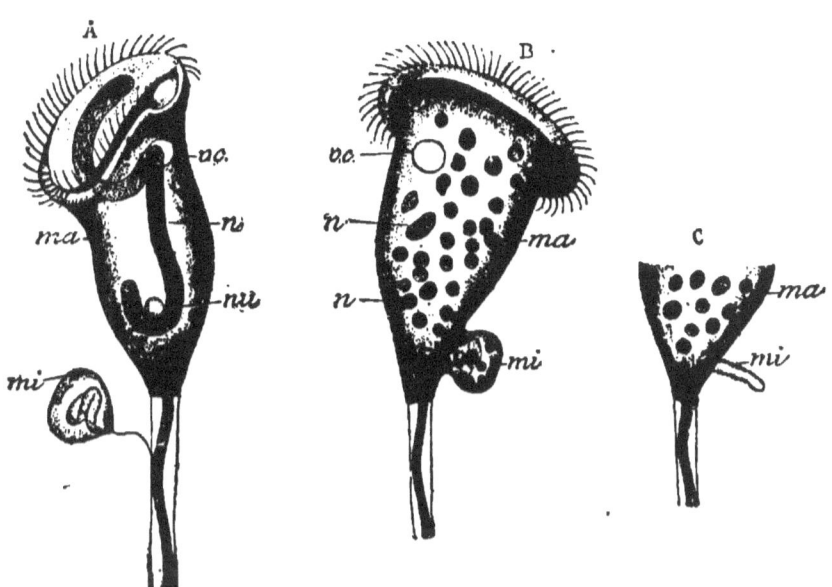

Fig. 7 *d.*—Gemmiform conjugation of the Vorticellinæ (*Carchesium poly-pinum*). *A*, first stage: the microgonidium *mi* has fastened itself by a filament upon the peduncle of the macrogonidium. *B*, a more advanced stage: the microgonidium has fastened itself directly upon the body of the macrogonidium, and its substance begins to penetrate into the latter. In both individuals the nucleus has separated into small rounded fragments, and in the microgonidium are seen the two striated segments resulting from the division of its nucleolus. *C*, last stage of the conjugation. The microgonidium, completely void of its contents, remains attached to the body of the macrogonidium under the form of a minute hollow tube, which in the end drops away,—*ma*, macrogonidium; *mi*, microgonidium; *n*, nucleus; *nu*, nucleolus; *v.c.*, contractile vesicle. (Figures from M. Balbiani.)

EXPLANATION OF FIG. 7 E.

Fig. 7 e.—Conjugation of the *Chilodon cucullulus*. *A*, beginning of conjugation; *b*, mouth; *n*, nucleus; *nu*, nucleolus; *v. c.*, multiple contractile vesicles. *B*, division of the nucleolus into two segments, *nu'*, *nu'*; the nucleus *n* begins to show signs of regression. *C*, each of the two individuals in conjugation contains two nucleolar segments, brought near together, of which one probably comes from the individual opposite by course of exchange, and will fuse with the segment not exchanged, to form a compound segment (Maupas). *D*, division of the segment into two portions which grow to unequal sizes; the larger, *nn*, will become the new nucleus, the smaller, the nucleolus of the new formation, *nun*. *E*, the old nucleus, *n*, reduced to a small pale and rumpled mass, is replaced by the new nucleus *nn*, near by which is seen the new nucleolus *nun*. (Figures from M. Balbiani.)

In the *Chilodon cucullulus* the following is the series of phenomena presented (the row of figures given opposite, have been procured through the kindness of M. Balbiani, and will serve to illustrate our description): The figure *A* shows the beginning of conjugation; each animal is shown with its mouth (*b*), its multiple contractile vesicles (*v. c.*), its nucleus (*n*), and its nucleolus (*nu*); the nucleolus will become the main seat of the modifications effected in fecundation. In the figure *B* the principal change pertains to the nucleolus; in each of the two animals, the nucleolus has moved away from the nucleus and has begun to break apart into two segments; the nucleus commences to show signs of regression. Between the figure *B* and the figure *C* phenomena take place of the highest importance, but which are still the subject of dispute. The following appears from present investigations to be the most probable: the two animals in conjugation exchange with each other one of the capsules produced by the division of the nucleolus, so that when we come to the phase of the process represented in figure *C*, we find an animal which contains, besides its nucleus, two nucleolar segments in immediate proximity (*nu' nu'*); it was the same in figure *B;* each animal already possessed two nucleolar segments, but these segments were obtained from the division of the nucleolus properly belonging to the animal itself, while in figure *C*, in consequence of an exchange effected, one of the nucleolar segments belongs originally to each animal and the other comes from its mate.

M. Balbiani, who made the first observations upon these phenomena of a nature so delicate and complex, originally supposed that the two adjacent nucleolar segments, which have been represented in the figure *C*, were produced by the longitudinal division of the nucleolus exchanged between the two animals in conjugation.

M. Maupas has recently proposed a different explanation, which seems to be further corroborated by the very figure given twenty years previously by M. Balbiani. According to M. Maupas the segment exchanged proceeds to fuse with the segment not exchanged, in order to form a compound segment; the two contiguous segments, seen in figure *C*, would not, therefore, be the result of the division of one segment solely, but the first stage of the conjugation of two elements having different origins. A fact which apparently argues in favor of this opinion, is the aspect presented by the two segments; if they proceed from a division, we would find there certain phenomena of caryokinesis, which were furthermore completely unknown at the time when M. Balbiani made his first observations.

However this may be, it is seen by figure *C* that the regression of the old nucleus (*n*) is sharply marked.

In the figure *D*, the two nucleolar segments have fused together and have formed a compound segment, which segmentates in its own turn; the two new products of that segmentation grow to unequal sizes; the largest capsule attains a size of forty thousandths of a mm.; it is this that forms the new nucleus of the *Chilodon.* The second capsule shrinks and becomes compressed, it takes its place beside the first one and becomes the new nucleolus.

The figure *E* represents the last stage of the phenomenon; the animal is in possession of its new nucleus and its new nucleolus;

the old nucleus is reduced to a small pale and rumpled mass and will shortly disappear.

To recapitulate, then, if the opinion of M. Maupas (who did not study this species, but like ones) be accepted, the nucleolus divides into two capsules: the one, playing the part of a male element, is exchanged between the two animals in conjugation, and proceeds to fuse with one of the capsules derived from the division of the nucleolus of the other animal; the other capsule, which acts the part of a female element fuses in the same way with the male element coming from the other animal. The result of that fusion is a compound capsule which, undergoing a process of division, produces the new nucleus and the new nucleolus of the animal fecundated.

ADDENDA.

NOTES, References, Authorities, etc., omitted in the text:

Page 1, line 16. The doctrine of unicellularity in regard to the Infusoria has been upheld by Sibold and Kölliker; the majority of naturalists have conceded it.

Page 9, line 21, et seq., vide *Pflüger's Arch.*, Vol. XXIII, 1880.

Page 10, line 4, vide Rouget, *Revue Scientifique.* March 15, 1884.

Page 10, line 13, vide *Annales des Sciences Naturelles*, 1835, Vol. IV, pp. 348 and 361.

Page 12, line 29, et seq., vide *Arbeiten aus dem zoölog. Institut in Würzburg*, herausgegeben von Prof. C. Semper, Vol. I. p. 9, 1872.

Page 15, lines 12 and 13, vide *Morphologisches Jahrbuch*, Vol. X. 1885, p. 534.

Page 16, line 25, vide *Pflüger's Arch.*, 1876.

Page 19, line 28, vide Balbiani, *Lecons sur les Sporozoaires*.

Page 20, line 6. By protoplasm in this connection is understood the entire cellular body; the distinction of function between the protoplasm properly so called, and the nucleus, is established later on in the essay.

Page 26, lines 9 and 10, vide *Comptes rendus de l'Acad. des Sciences*, Nov. 2, 1886, No. 18.

Page 29, line 26, vide *Bot. Zeitung*, 1881, 1883, 1884, 1886.

Page 46, last line, vide E. Maupas, *Etude des Infusoires ciliés*, *Arch. de zoöl. expér.*, 1883, No. 4.

Page 58, lines 30 and 31, vide Henneguy, *Sur la reproduction du Volvox dioique. Acad. des Sciences*, July 24, 1876.